DATE DUE			
May 6 '73			

A
SHORT HISTORY
OF ANATOMY
FROM
THE GREEKS
TO HARVEY

Title page of the first edition of Vesalius *De Fabrica corporis humani*; Basel 1543.

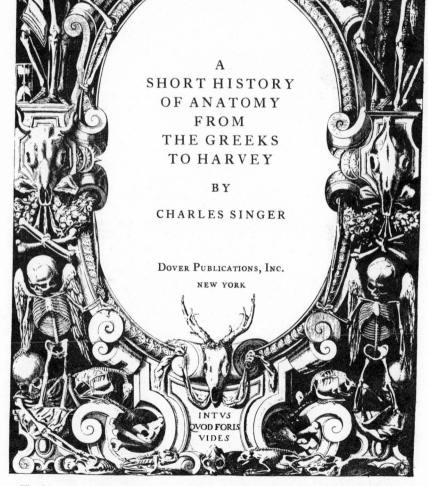

A
SHORT HISTORY
OF ANATOMY
FROM
THE GREEKS
TO HARVEY

BY

CHARLES SINGER

DOVER PUBLICATIONS, INC.

NEW YORK

INTVS
QVOD FORIS
VIDES

The historiated Title page of this volume is taken from GIULIO CASSERIO, *De vocis auditusque organis, historia anatomica*, 1601.

This Dover edition, first published in 1957, is an unabridged republication of the first edition with a new Preface by the author, originally published under the title *The Evolution of Anatomy.*

Standard Book Number: 486-20389-1
Library of Congress Catalog Card Number: 57-4484

Manufactured in the United States of America
Dover Publications, Inc.
180 Varick Street
New York, N.Y. 10014

For out of olde feldes, as men seith
Cometh al this new corn fro yeer to yere;
And out of olde bokes, in good feith,
Cometh al this new science that men lere.

CHAUCER: *The Parlement of Foules,* lines 22-25.

PREFACE TO SECOND EDITION

This work has been out of print for many years. I long hoped that someone more competent than I would cover the ground more completely. This has not happened and, since the demand for the book has continued, it is here revised with a number of minor adjustments. I would express the hope that this new edition may rouse sufficient practical interest in the subject to induce someone to undertake a more comprehensive work. The literature in English is now more copious than it was when the first edition was published over thirty years ago. Works by some of the leading figures in the history of anatomy — Vesalius, Galen, Coiter, Mondino, Leonardo — are now available in English translation, and I am in hope that this new edition may turn my readers' attention at least to these.

CHARLES SINGER

"Kilmarth"
Par, Cornwall
26th.May 1956

PREFACE

I HAVE to return thanks first to the College of Physicians and its President who invited me to give the Fitzpatrick Lectures and next to three colleagues who helped me in their preparation.

Association with Professor Platt has been a very great stimulus ; to him I owe inestimable help in acquiring such knowledge of ancient Medicine as I may possess ; his great sympathy with Science and with its History has been a source of inspiration to many, but I have been unusually privileged in being permitted to draw on that store of exact knowledge of Greek on which his varied learning was based as on a rock. He read this book in manuscript and it is a very great grief to me that he has not lived to see it printed. The first two Chapters owe a debt to him on almost every page. For the last two chapters I have received similar generous help from Emeritus Professor Sir George Thane, whose unrivalled knowledge of every department of Anatomy and of its literature, has been unweariedly placed at my disposal. I have enjoyed every advantage afforded by the magnificent Anatomical Institute of Professor Elliot Smith from the time that it was built, and especially I have had the fullest use of its very fine collection of old anatomical books deposited there. This collection was bequeathed to University College by William Sharpey, who once held the chair that has since been occupied by Professors Sir George Thane and Elliot Smith. My Fitzpatrick Lectures were themselves based on a course given to some of Professor Elliot Smith's more advanced pupils, and I have to thank him for help at many points and for sympathy and understanding throughout.

A number of other colleagues have aided me in various directions. Mr. T. L. Poulton, artist to the Anatomical Department at University College, has made drawings for me which are reproduced in Figs. 24, 26, and 28. The other drawings, maps, and tracings are the work of my secretary, Miss E. Biden, or of myself. Dr. H. H. Woollard, Assistant Professor of Anatomy at University College, London, has read the book in proof and corrected points of nomenclature. Dr. H. A. Harris, Senior Demonstrator of Anatomy at

University College, London, has advised me on certain other anatomical matters. Dr. C. F. Sonntag, Demonstrator of Anatomy at University College, London, has helped me with information on simian Anatomy. He has also provided me with the hand of a Barbary ape, the dissection of which (Fig. 27) has been the work of Dr. Kozinski, of the University of Wilna, who is at present engaged in research at University College. Miss Margaret Murray, Reader in Egyptology at University College, London, drew my attention to the Egyptian material reproduced in Figs. 5, 6, 7, and 8.

Mr. W. G. Spencer, O.B.E., has most generously placed at my disposal the blocks from which are printed the full page figures from Vesalius. I have also to thank Messrs. Bell & Son, the Oxford University Press, and Messrs. John Wright, of Bristol, for the use of a number of clichées. Sir Arthur Evans drew my attention to the arm of Cretan origin shown in Fig. 4 ; he also kindly gave me the photograph from which that figure has been drawn. Professor K. Sudhoff has been good enough to supply me with the photographs from which Plate XIV is taken.

Lastly, I have to thank the Honorary Librarians of the Royal College of Physicians of London and the Royal Society of Medicine for unusual facilities for the study and photography of some of the treasures under their charge.

The main task in preparing this account of the History of Anatomy has been the investigation of sources. I do not think I have referred to any book or manuscript without having myself examined either the original or a direct photograph or facsimile. Suggestions, however, have been derived from the few general accounts of the History of Anatomy that have so far appeared. Of these the most useful are still the detailed but somewhat confused work of Baron Portal, which appeared as long ago as 1770, and the unfinished but very creditable attempt of Lauth in 1815. There is also the fine bibliography of Albrecht von Haller dating from the years 1774-7. The inaccurate treatise of Burggraeve, which had run through three editions by 1880, I have found to be the source of many errors that have since gained currency. I have found it useless for practical purposes. A great number of facts concerning the History of Anatomy are conveniently catalogued by von Töply in the second volume of Puschmann, Neuburger, and Pagel's *Handbuch der*

Geschichte der Medizin, which appeared in 1903. There are
naturally many memoirs dealing with special aspects of
Anatomy or on special periods of anatomical study, which
I have read for the purpose of this book. Among such works
are those of Albertotti, F. Baker, Cervetto, Choulant, F. J.
Cole, Corner, Corradi, Del Gaizo, Daremberg, Duhem, Duval,
Ferckel, Fonahn, A. Forster, M. Foster, Garrison, Hirschberg,
Holländer, Holl, Holma, Hyrtl, Ilberg, Jackschath, Jastrow,
Klebs, Locy, Littré, Macalister, MacMurrich, Martinotti, Marx,
Medici, Milne, Moores-Ball, Miss M. Murray, Neuberger,
Nicaise, Pagel, Petrie, Pilcher, Piumati, D'Arcy Power,
Puschmann, Redeker, Regnault, Roth, Schöne, Seailles,
Seidel, C. G. Seligmann, W. G. Spencer, Spielmann, Stieda,
Streeter, Sudhoff, Tarrasch, De Toni, von Töply, Walsh,
Walston, Washburn, Wegener, Wellmann, Weindler,
Weyermann, Wickersheimer, W. Wright, and many others.
There are also a few sketches of the subject as a whole, either
in the form of Introductions to anatomical treatises or short
independent pamphlets or lectures. Perhaps the best of the
latter class was one printed just a century ago. It was delivered
by Matthew Baillie in 1785 as an *Introductory Lecture* to
the course in Anatomy at the Great Windmill Street School
founded by Baillie's uncle, William Hunter. This lecture
will be found in a rare little book entitled *Lectures and
Observations on Medicine by the late Matthew Baillie, M.D.,*
published posthumously by Baillie's executors and privately
issued in 1825 in a very small edition.

In composing this short History of Anatomy I have been
concerned to keep it within compassable limits. Unprogressive
anatomical movements and periods are therefore but lightly
touched upon, attention being concentrated on the line of
definite advance. It appeared both unnecessary and undesirable
to make any division between Physiology and Anatomy, at
least in the period under consideration. As the narrative
of the little book ends early in the seventeenth century, no
mention has been made of the beginnings of Iatrochemistry
in the persons of Paracelsus, van Helmont, and their followers,
since the movement they represent did not become important
until the second part of the seventeenth century. The volume
closes with Harvey. The new physiological movement,
together with later anatomical developments will be treated
in a separate work.

In discussing individual writers I have made no attempt
at consistence in the treatment of names. I have sometimes
spoken of a man by the name his mother called him, as with
Mondino ; sometimes I have used the Latinized form as with
Vesalius ; sometimes I have used the conventional title as
with *Galen*. In all this I have been guided by what I conceive
to be the reader's ease.

" Vivitur ingenio, caetera mortis erunt," *It is his genius
that yet walks the earth ; all else of him may go down into
silence,* is the motto which Vesalius himself has chosen for
the most beautiful of all his figures (Fig. 103). Let him be
taken at his word ! I have sought to treat him and the great
men who went before and after him as they would be treated.
Prompted alike by personal inclination, by necessity for
brevity, and by the suggestion of Vesalius I have considered
only the actual contributions to knowledge that these men
have made, seeking to treat the History of Anatomy as a
secular conversation between great minds, a debate of men
of genius continued through the ages. If the writing of
History cannot establish such continuity of ideas it can work
nothing effectual. The reader who seeks information concerning the ancestry after the flesh or the details of the
domestic life of these anatomical heroes, should turn to
one of the several large dictionaries of medical biography.

With a view to the greatest possible compression, references,
quotations, and bibliography are omitted from this preliminary
sketch. They must await a more extensive work on the
History of Anatomy. It has, however, been necessary to
deal with a number of contentious matters, especially in
Chapters I and III. On such points I have often ventured
to take my own line. This is done not in any dogmatic
spirit but simply for the sake of brevity. I trust, especially,
the reader will not regard me as too arbitrary in my dating
and treatment of the documents of the Hippocratic Collection.
Such topics need lengthy discussion of a kind that appeals
to few. It therefore seemed best to leave aside all documentation from such a work as this. It may be that there are readers
who will be not ungrateful for the postponement.

CHARLES SINGER.

ANATOMICAL INSTITUTE,
    UNIVERSITY COLLEGE, LONDON.
        May 26, 1925.

CONTENTS

LIST OF PLATES

facing page

A
SHORT HISTORY
OF ANATOMY
FROM
THE GREEKS
TO HARVEY

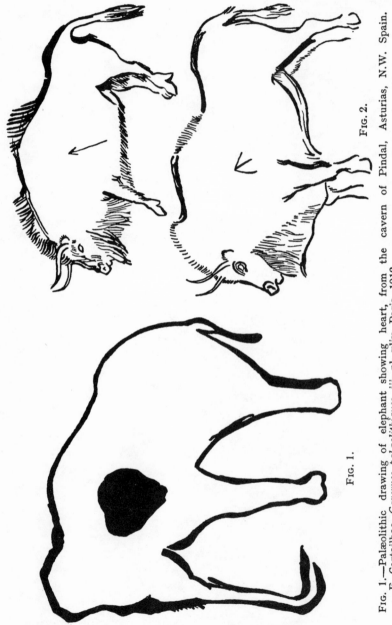

Fig. 1.

Fig. 2.

Fig. 1.—Palaeolithic drawing of elephant showing heart, from the cavern of Pindal, Asturias, N.W. Spain. E. Cartailhac, *Cavernes palæolithiques*, iii, pl. xliv, Paris, 1912.

Fig. 2.—Magdalenian drawings of bison with arrows embedded in the heart, from the cavern of Niaux on the Ariège, S. France. E. Cartailhac and H. Breuil, *L'Anthropologie*, xix, p. 15, Paris, 1908.

THE EVOLUTION OF ANATOMY

I

THE GREEKS TO 50 B.C.

§ 1   *The Prescientific Stage*

THERE is, in a sense, an *anatomical instinct*. To reach its first beginnings we should have to carry our search far back indeed. Nor could we safely stop even with humankind. We might discern rudiments of it in the neck-breaking clutch of the tiger or in the accurate puncturing by the

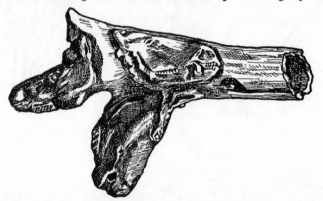

FIG. 3.—Late Magdalenian carving of three horse heads in reindeer horn from the great cavern of Mas d'Azil, Department of Ariège, S. France.   The head below is carved from life.   From the head above the tissues have been removed and hardly more than the skull is left.   The head that projects to the left has been skinned and the contours of the muscles are seen.   *Revue anthropologique,* 1909, p. 396.

ichneumon-fly of the ventral ganglia of its victim. Or, if we turn from such acts to the conscious achievements of our own species, we may yet discern in the lore of the hunter or the craft of the butcher an accurate grasp of anatomical fact, albeit adapted only to certain ends and confined within a restricted field. The palæolithic bowman well knew where to find the

heart of his victim (Fig. 1), and he has portrayed it trans-
fixed with arrows on the walls of his shelter (Fig. 2).   The
artist who worked in the cavern at Mas d'Azil has left us
many memorials of his skill and knowledge.   Among them
are carvings of the skulls of horses and even a representation
of a dissection of the horse's head exhibiting the contour of
the surface muscles (Fig. 3).

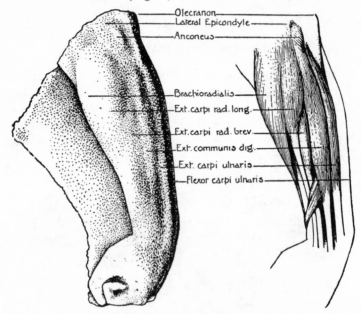

FIG. 4.—Clay model of forearm from Cnossus in Crete of about
A.D. 1500. (Late Minoan I).   The contours of the muscles are
so well shown as to suggest anatomical knowledge.   From a
photograph kindly supplied by Sir Arthur Evans.   A modern
dissection has been placed by it for comparison.   The drawings
are the work of Mr. T. L. Poulton, artist to the Anatomical
Department at University College.

We shall not, however, begin on such levels as these.   We
shall turn rather to Anatomy in the scientific stage when
definite anatomical conceptions of a more generalized character
are consciously formulated and consciously accumulated.
At a certain level in human development such knowledge or
such tradition comes to be set forth for its own sake.   It begins
to be expounded for the satisfaction of human curiosity, and

thus a new motive is added to the mere desire to meet the
needs of some art or craft. Then, and then only, can we truly
say that we have to do with Science. The scientific stage is
reached very late in human development. As we stand before
the vista of Human History, we shall not have to look back
very far if we seek only the origin of Anatomy as a Science.

Our anatomical tradition, like that of every other department of rational investigation, goes back to the Greeks. From
them the methods, the applications, and even the very
nomenclature of our anatomical discipline are more or less
directly derived.

It must not be too hastily assumed, however, that the
Greeks themselves had no anatomical forerunners. Before
the coming of the Greeks the Ægean area was inhabited
by the so-called Minoan peoples. Their culture has been
revealed during the last generation at its metropolis in Crete.
There many characteristic works of art of a high order have
come to light. Not a few present us with a close study of
surface contours of the body of man (Fig. 4), comparable to
those by the cave artist of the creatures on which he preyed.
Judgment must, however, be suspended as to whether this
knowledge of the Minoan artist was controlled by any
anatomical tradition.

The Greeks, however, were little dependent on their
Ægean predecessors. Between the ages in which the Minoan
culture flourished and that in which the Hellenic civilization
came to flower there yawns a chasm of many centuries. On
the ancient civilization of the Euphrates and the Nile
the Greeks could more readily draw. From an early date
Greek travellers and traders penetrated into Egypt and could
there peruse written records of great antiquity. We find,
in fact, that certain Egyptian medical papyri set forth surgical
procedure demanding considerable anatomical knowledge.
Other Egyptian documents give accounts of the structure of
the human body, which, however bizarre they may be, yet
show the conviction that the human body has a definite
and ascertainable structure. We are perhaps unfortunate in
the character of the medical material that Egypt has so far
yielded. There is, indeed, evidence that Egyptian knowledge

of anatomy may have been less incomplete and unscientific than the discovered documents have as yet suggested. Thus the traditional form of the womb, as it appears in mythological contexts (Figs. 5 and 6), suggests some genuine access to

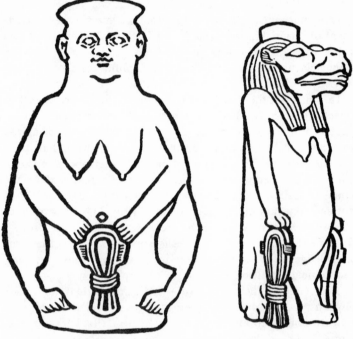

FIG. 5.   FIG. 6.

Figures representing Taurt, an Egyptian goddess associated with childbirth, redrawn from C. G. Seligmann and M. A. Murray, *Man*, xi, 113, London, 1911.

FIG. 5.—An Alabaster vase of the XVIIIth dynasty showing the human-headed form of the goddess. She clasps an object resembling the hieroglyphic *sa* sign (see Fig. 7), holding it in a position which suggests the uterus.

FIG. 6.—The usual hippopotamus-headed form of the goddess holding a *sa*-like object in each hand.

anatomical sources. Taurt, the Egyptian hippopotamus-headed goddess of childbirth, is almost invariably represented as carrying this sign either in front (Fig. 5) or at the side (Fig. 6). It occurs as a hieroglyphic from the

3rd Dynasty (about 2900 B.C.) onward (Fig. 7). Even in this conventionalized form it is, in fact, more like the real object than the representations of the mediæval anatomists, prepared at a period when dissection was regularly practised. Again the heart and trachea are frequently represented in

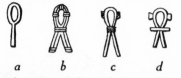

a b c d

FIG. 7.—A series of representations of the *sa* sign derived from the form of the uterus. From C. G. Seligmann and M. A. Murray, *Man*, xi, 115, London, 1911.
a, Dynasty III. b, Dynasty XII. c, Dynasty XVIII. d, Ptolemaic.

Egyptian amulets and a conventionalized representation of these organs has passed into the hieroglyphic system (Fig. 8). Thus there was certainly some anatomical tradition in Egypt when that country began to be penetrated by the Greeks. There are points in the history of Anatomy among the Greeks which suggest contact with Egyptian ideas.

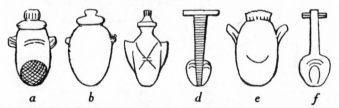

a b d e f

FIG. 8.—Egyptian conventional representations of the heart (a, b, c, and e) and of the trachea and lungs (d and f). a, b, and c are amulets and are from W. M. F. Petrie, *Amulets*, London, 1914. d, e, and f are hieroglyphs. d and e are from F. Ll. Griffith, *A collection of Hieroglyphics, a contribution to the History of Egyptian writing*, London, 1898. f is from N. de G. Davies, *The Mastaba of Ptahhetep and Akhethetep at Saqquareh*, London, 1900.

Similarly, from the early Mesopotamian civilization, accounts of surgical procedure have reached us that seem to involve anatomical knowledge. The code of Hammurabi (*c.* 1950 B.C.), perhaps the Amraphel of Scripture history (Genesis xiv, 1), tells of a complex surgical tradition. There springs also to the mind that masterpiece of Assyrian art, the dying lioness, from the palace of Assurbanipal (reigned 668–626 B.C.).

The stricken creature, with her spinal cord severed, crawls upon her tormentors, snarling furiously, dragging her paralysed hind limbs (Plate I). We recall the experiments on the spinal cord made by Galen (see p. 60) eight hundred years later in which such observations were brought at last under

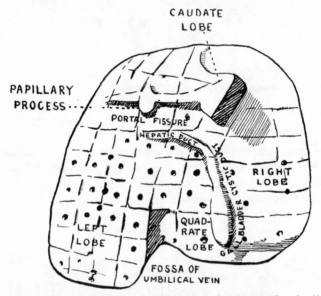

FIG. 9.—Clay model of sheep's liver used for instruction in liver divination in a Babylonian temple school. The model is covered with cuneiform writing which, for the sake of simplicity, has been omitted here. The inscription, which fixes the date of the object at about 2000 B.C., furnishes the prognostications for peculiarities noted at the parts of the liver indicated by holes. The model is therefore a diagram to explain an omen text in which the peculiarities in question were registered together with the interpretation to be attached to them. The lobes of the liver, the portal fissure, the gall bladder, and the cystic and part of the hepatic duct are shown. Technical names are given to these and other parts. The lines indicate conventionally the markings due to the tracing on the surface of the subsidiary ducts that collect bile into the main duct. (British Museum.) See M. Jastrow, *Proceedings of the Royal Society of Medicine* (*Historical Section*), vii, p. 109, London, 1914.

more scientific control. From Mesopotamia we have, moreover, clay models of the liver of the sheep used for divination (Fig. 9) at an early date. These suggest a knowledge of animal anatomy not inferior to the earlier Greek material.

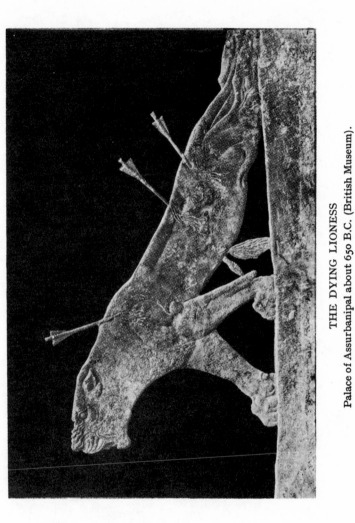

PLATE I

THE DYING LIONESS

Palace of Assurbanipal about 650 B.C. (British Museum).

[*face p.* 8

The Greeks certainly drew some of their medical lore from Mesopotamia, deriving therefrom, for instance, the names of certain drugs. It is thus not unlikely that they may also have taken anatomical hints from the same source. Their scientific debt to Egypt they were ready—perhaps too ready—to acknowledge. The relation of the Minoan culture to that of Hellas is not yet clear, but there must have been much Minoan blood among the mixed peoples whom we call Greeks. Crete, Babylon, and Egypt, however, yield us but broken lights, and for practical purposes our knowledge begins with the Greeks themselves.

§ 2   *The Schools of Sicily, Ionia, and Cos.*   550 B.C.–400 B.C.

We need not delay with the pre-scientific stage of Greek civilization. Material of medical and even of anatomical interest has been elicited from the writings of Homer (about c900 B.C.) and of Hesiod (about 750 B.C.), and from the early Greek monuments. These, however, are on a cultural level below those of Egypt and Babylon. We shall begin with the earliest records of actual anatomical observations, and these are to be found in the fragments of the writings of Alcmæon (about 500 B.C.), a native of the Greek colony of Croton, in southern Italy. Alcmaeon began to construct a positive basis for medical science by the practice of dissection of animals. He discovered the optic nerves, and the tubes called in after ages by the name of Eustachius (p. 135). He even extended his researches to Embryology, describing the head of the fœtus as the first part to be developed—a justifiable deduction from the appearances. Curiosity excited by him as to the distribution of the vessels led his followers, Acron (about 480 B.C.), Pausanias (about 480 B.C.), and later Philistion of Locroi (about 390 B.C.), the contemporary of Plato, to make anatomical investigations.

Very important among these early Greek writers for his influence on later thought was Empedocles (about 480 B.C.) of Acragas in Sicily. His view that the blood is the seat of the *innate heat* he took from folk belief—" the blood is the life." It recurs in many later authors. His teaching led to a belief in the heart as the centre of the vascular system, and the chief organ of the *pneuma* which was distributed by blood-vessels.   This pneuma was equivalent to both soul and life, but it was something more. It was identified with air and breath, and the pneuma could be seen to rise as a shimmering steam from the shed blood of the sacrificial victim—for was not the blood its natural home ? There was a pneuma, too, that interpenetrated the Universe around us and gave it those qualities of life that it was felt to possess. Anaximenes (about 560 B.C.), an Ionian predecessor of Empedocles, had defined these functions of pneuma with his phrase : " As our soul, being air, sustains us, so pneuma and air pervade the whole World."   It was, however, the speculations of Empedocles himself that came to be regarded as the basis of the Pneumatic School in Medicine, which had later very important developments.

The views of Empedocles, and especially his doctrine that regarded the heart as the main site of the pneuma, were rejected by the Coan school—a group of medical writers who came into prominence in Western Asia Minor during the fifth century—whose works have since been fathered on Hippo-crates of Cos (Plate IV).   Empedoclean doctrine, however, was not without influence on Ionia, which at this time led the van of Greek thought. Thus Diogenes of Apollonia, a late fifth century writer who was approximately contemporary with Hippocrates himself, was profoundly influenced by the Pneumatic School. Diogenes made an investigation of the blood vessels (Fig. 10), and it is clear that the interest in the subject had become widespread at this time.

Another work of the late fifth century and contemporary with this Diogenes is the peculiar treatise *On regimen*, which finds a place in the Hippocratic Collection.   It is strongly under the influence of the thought of Heracleitus (about 540–475, Plate II), and contains many points of view which reappear in later philosophy. All animals according to it are formed of fire and water, nothing is born and nothing dies, but there is a perpetual and eternal revolution of things, so that change itself is the only reality. Man's nature is but a parallel to that of the universal nature, and the arts of man are but an

PLATE II

PLATO (427-347 B.C.)
Statue in Vatican. Work of First Century A.D.
copied from original of about 380 B.C.

HERACLEITUS OF EPHESUS (c. 540-475 B.C.)
Statue in Museum at Candia from Agora at Gortyna. Work of
Second Century A.D. copied from original of Fifth
Century B.C.

Inset: Ephesian coin of Fifth Century B.C. showing similar figure.

[face p. 12

bearing on the course of later biological thought, that in the
fœtus all parts are formed simultaneously. On the proportion
of fire and water in the body all depends—sex, temper,
temperament, intellect.

We shall not dwell on the shadowy and elusive figure of
Hippocrates of Cos (about 400 B.C., Plate V). There is no
satisfactory evidence as to which works, if any, in the so-called
" Hippocratic Collection " are really from his hand, though
there is abundant evidence that the works in that Collection
are by *many* hands. Among the sections of the Hippocratic
Collection that are anatomically more interesting, however,
is the treatise *On the sacred disease* which may be referred with
reasonable confidence to about the year 400 B.C. This work is
of very great importance for the history of Philosophy as an
early attempt to set forth a rational view of the Universe
based on the conception of Natural Law.

The brain of man is represented in the *Sacred disease* as
resembling that of all other animals in being cleft into two
symmetrical halves by a vertical membrane. To the brain
there come many blood-vessels, some slender, but two stout.
One of these stouter vessels is said to come from the liver and
the other from the spleen. This extraordinary statement may
be an alteration of an original which said that one came from
the side corresponding to the liver and the other *from the side
of* the spleen. The description of the vessels is indeed confused.
We read, however, of a great blood-vessel that passes upward
under the collar bone by the side of the neck, is visible there
beneath the skin, and finally buries itself as it reaches the ear,
where it divides into branches In the course of the discussion,
the writer tells us that he has cut open the skulls of goats to
examine the brain. The arteries are said to contain air, an
idea gained from their emptiness in dead animals. At certain
points the work bears a resemblance to passages in one of the
Egyptian medical papyri.

A work kindred to the *Sacred disease*, and one belonging to
the same school, of similar date, and not improbably by the
same author, is the famous treatise *On airs, waters, and places.*
That work contains a crude but not unscientific attempt to
classify the races of men by their physical characteristics and

imitation or reflex of the natural arts, or, again, of the bodily
functions. The soul, a mixture of fire and water, consumes

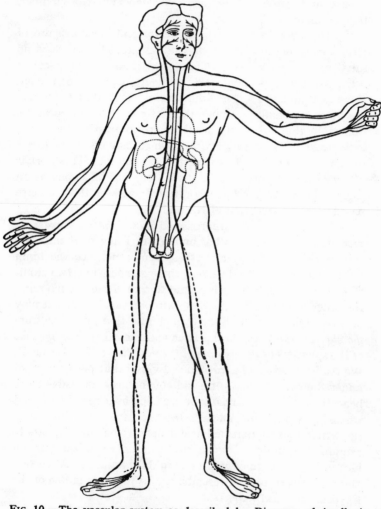

FIG. 10.—The vascular system as described by Diogenes of Apollonia
about 400 B.C., slightly modified from E. Krause, *Diogenes von
Apollonia*, Posen, 1909.

itself in infancy and old age and increases during adult life.
Here, too, we meet with that strange doctrine, not without

the lands in which they dwell. Its author thus ranks along with Herodotus (about 484–425 B.C.) as one of the fathers of the science of Anthropology. The view of the inhabited world that we encounter in the *Airs, waters, and places* is confined to Greece and its islands, Southern Russia, Asia Minor, and Egypt, and is more limited than that of Herodotus (Fig. 11). There is no knowledge of Italy.

A very different work is the *Wounds of the head*, also of about 400 B.C., and also included in the Hippocratic Collection. The operation of trephining is admirably described, and the general tone is entirely scientific. The operation itself was well known from an extremely early date. It was frequently

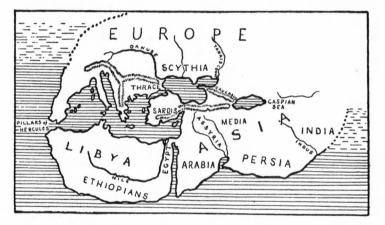

FIG. 11.—The world according to Herodotus.

practised with stone implements in prehistoric times and is still resorted to by many savage tribes. In the Hippocratic work, however, we have both a technique and, above all, a scientific interest that is vastly in advance of the powers of Neolithic man. Thus the different types of skull are described, and variations in the sutures are pointed out in some detail though without exactness. Some of the errors may be due to changes and dislocations of the text, for the work is manifestly from the hand of a practical surgeon, well accustomed to treat cases of cranial injury. It is the source of the doctrine of *fracture by contrecoup*, which held an important place in Surgery until the present generation. This composition bears distinct analogies at some points to an Egyptian medical papyrus that has recently come to light.

Little later than the *Wounds of the head* is the work of Polybus (about 390), the son-in-law of Hippocrates, *On the nature of man*. It contains a clear statement of that doctrine of the four humours which was to play a paramount rôle in medical thought for the next two millennia. These humours— Blood, Phlegm (*pituita*), Black Bile (*melancholia*) and Yellow Bile (*cholē*)—together make up the living body in the same way as the four elements with which they correspond—Fire, Water, Earth, and Air—make up non-living matter. The four elements themselves are related to the four qualities in later literature (Fig. 16).

The question has sometimes been raised whether the great masterpieces of Greek sculpture, with their very close study of surface musculature, do not imply some anatomical knowledge. The answer must be given in the negative. During the great period of Greek art, the fifth century, there is no evidence that dissection of the human body had yet been practised. There can be no doubt that the muscular contours, as represented in works of the period, were studied from the living and not from the dead model. Occasionally the contour of a muscle is represented to which modern works on Anatomy for artists have failed to draw attention. An interesting case is that of the *Pectineus* muscle, shown in a statue from the Argive Heræum of about 450 B.C. Since the statue was discovered fifty years ago it has been demonstrated that in certain positions and under certain conditions this muscle can, in fact, be detected in the living subject. In later Greek sculpture, from about 200 B.C. onward, it is, however, possible that the artists were working on a real anatomical tradition derived from dissection. At that period, however, we have, as we shall see, independent records of the practice of dissection.

A few vase paintings have come down to us in which medical scenes are represented. In these we might expect a specially exact study of surface anatomy. One, of the early fifth century, is the work of a known and admirable artist,

Euphronius. The treatment of the hands is very fine, but the anatomy as a whole is disappointing. Another is of a clinic of the end of the fifth century. It shows a physician in the act of bleeding from a vein at the bend of the elbow. Behind him stands an achondroplastic dwarf admirably portrayed (Plate III). Here, too, we may mention the interesting Samian metrological relief of the middle of the fifth century. It represents an exact study of human proportions (Plate IV), and shows that the " Canon of Proportion " was already fixed, as we should expect, at a time when Greek art was at its zenith.

§ 3 *The Early Athenian Period, about* 400 B.C.–350 B.C. *Plato, Diocles*

From about 400 B.C. Athens becomes the main centre of anatomical activity. The controlling factor in the development of thought in this period is the great intellectual revolution instituted by Socrates (471–399 B.C.) and by his pupil Plato (427–367 B.C., Plate). The teaching of these great thinkers was not favourable to the development of physical investigation. This comes out notably in connexion with Anatomy. In the *Timaeus*, Plato sets forth an entirely fanciful scheme of the human body. The work was early translated into Latin and deeply influenced the Middle Ages. It is interesting as exhibiting a tendency to trace a parallel between the outer world, or, as it was afterwards called, the *Macrocosm* [i.e. *Great World*], and man's body, afterwards called by contrast the *Microcosm* [i.e. *Small World*]. The world itself is represented as a living being, and all matter is endowed with life (*Hylozoism*). The doctrine of Macrocosm and Microcosm influenced the subsequent development of anatomical thought both profoundly and unfavourably (p. 65 and Plates VIII and XVI).

Scientifically more important than the *Timaeus* is a work in the Hippocratic Collection of about 370 B.C., *On generation*. The treatise is peculiarly interesting as propounding a doctrine of *pangenesis* to account for the phenomena of heredity. That doctrine bears most striking analogy to the theory enunciated by Charles Darwin (1809–82) in his *Variations of Animals and Plants under Domestication* (1868). The writer of the work *On generation* believes that channels pass from all organs to the brain, thence to the spinal marrow, thence to the kidney, and finally to the generative organs. According to this work, acquired characters are inherited. The embryo develops and breathes by material transmitted through the umbilical cord. There is a noteworthy description of the *membrana mucosa uteri*. The human embryo is compared with that of the chick.

Diocles of Carystus in Euboea practised in Athens about the middle of the fourth pre-Christian century. He knew of some, at least, of the writings that are to be found in the Hippocratic Collection, but derived his inspiration from the West—Sicily and Italy—rather than from Ionia or Cos. Diocles was, however, an eclectic, and drew his opinions from many sources. Thus he adopted both the doctrine of the humours (cf. Polybus of the school of Cos, *On the nature of man*, p. 14) and of the innate heat (cf. Empedocles of the Sicilian school, pp. 9–10). He regarded the heart as the principal organ and seat of intelligence (cf. Aristotle, p. 19), and accepted " pneumatic " views (cf. Empedocles, p. 10). He developed embryological theory, holding, contrary to Aristotle, that the seed came from both sexes. Diocles claimed to have examined a fœtus of twenty-seven days and to have found traces of the head and of the spinal column. At forty days he was able to distinguish the form as human. His anatomical conclusions were based, to some slight extent, on human material, but it was chiefly animals that he actually dissected. He described the cotyledonous placenta. He wrote a book *On anatomy*, which has unfortunately disappeared, like his other writings.

The tract *On anatomy* that finds a place in the Hippocratic Collection is, perhaps, the earliest treatise devoted to the subject that we possess. In it is represented the standard of knowledge of the middle of the fourth pre-Christian century. It is unfortunately, however, only the merest sketch, and its text is corrupt.

Slightly later—perhaps of about 340 B.C.—is another member of the Hippocratic Collection, the treatise *On the heart*. This is our best representative of Athenian Anatomy

PLATE III

A GREEK CLINIC OF ABOUT 400 B.C.

From a vase in private possession at Paris. E. Pottier,
Fondation Eugène Piot, Monuments et Mémoires XIII, 149 Paris, 1906.

[*face p.* 16

of the period.   It was produced under the influence of such
" Western " writers as Alcmæon and Empedocles, and it seems
to be complete.   We cannot be certain whether or no it is
based on human dissection, but its author refers to the
anatomical similarity of man and animals.   The treatise *On the
heart* displays the doctrine of the *innate heat*, rejects the idea
that this mysterious entity resides in the blood, and elects
for the heart as its site.   Air, it is suggested, enters the heart
direct, and in the left ventricle of that organ some subtle
change of blood into spirit takes place.   It is there that the
intellect resides.   The work contains a description of the
auricles, of the auriculo-ventricular and semi-lunar valves, of
the *columnæ carneæ* and of the *chordæ tendineæ*.   An account
is given of an experiment for testing the competence of the
cardiac valves.   Very extraordinary to our modern ideas is
the statement—verified by experiment !—that, in drinking,
some of the fluid passes to the lungs ; yet this view is also
expressed in Plato's *Timaeus* and other early writings.

§ 4   *The Later Athenian Period, about* 350 B.C.–290 B.C .
Aristotle, Theophrastus

Aristotle (384–322 B.C., Plate IV), who was the son of a
physician, was the great codifier of ancient Science, and on him
all subsequent biological development, including that of
modern times, is surely based.   In his three great biological
works, the *History of animals*, the *Parts of animals*, and the
Generation of animals, he discusses many biological problems
current to this very day.   He laid the basis of the doctrine
of Organic Evolution in his teaching concerning the *scala
naturæ* (Fig. 12), he developed coherent theories of generation
and heredity, and he founded Comparative Anatomy.   It
may be taken as tolerably certain, however, that he never
dissected the human body.

Aristotle gave good descriptions of some organs regarded
from the standpoint of Comparative Anatomy.   These
descriptions he sometimes illustrated by drawings, the first
anatomical figures of which we have a record.   In some

cases these drawings can be restored with confidence, as, for instance, his representation of the male organs of generation (Fig. 13). He gave a description of the uterus the nomenclature of which has been retained, in more or less modified form, to our own time (Fig. 14). Among the best anatomical descriptions given by Aristotle is that of the stomach of the ruminant. Perhaps his most extraordinary anatomical feat is his account of the placental development of the dogfish *Mustelus lævis* (Fig. 15). This raised the admiration of the greatest modern morphologist, Johannes Müller (1807–58), and would in itself be sufficient to establish

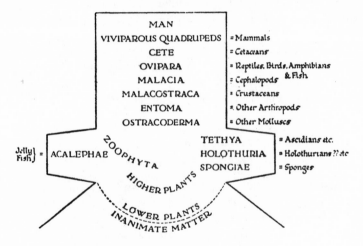

FIG. 12.—Scala naturæ of Aristotle.

the claim of Aristotle to a place in the front rank of observing naturalists.

Aristotle had paid special attention to the habits and structure and especially the breeding of fish. He knew that they were mainly oviparous, but occasionally viviparous, and he knew also of one instance among the Elasmobranch fishes (which he called *Selachia*), in which the development bore an analogy to that of placental mammals. This fact remained almost unnoticed until the nineteenth century, and it was its rediscovery that drew the attention of naturalists to the great value and interest of the Aristotelian biological masterpieces.

Something should be said of the anatomical errors of Aristotle. Most remarkable is his refusal to attach great

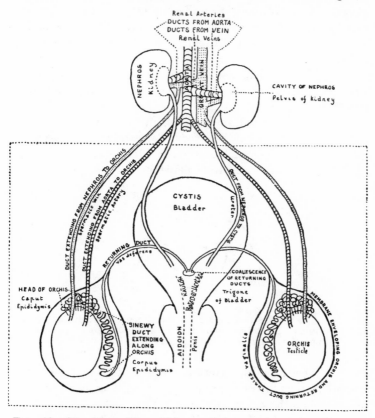

Fig. 13.—The Mammalian Urino-genital system as described by Aristotle. The part framed in a dotted rectangle is a restoration of a lost diagram prepared by Aristotle himself and described in his *Historia animalium*. The legends in capitals give the terms employed by Aristotle. Below these the modern equivalents are written in *italics*.

importance to the brain. Primacy he placed with the heart, where also was the seat of the intelligence. This was contrary to the view of the Hippocratic work *On the sacred disease* (about 400 B.C.), and contrary, too, to the view of most medical

writers of his day. It was contrary also to the popular view
voiced, for instance, by Aristophanes in his play *The clouds*
(about 400 B.C.), where we read of a man who had *concussion
of the brain*. Plato, too, in the *Timaeus* (about 380 B.C.),
placed the seat of thought and feeling in the brain. It is not
improbable that Aristotle had made experiments on the brain
and found it devoid of sensation. Hence his view, in opposition
to current belief, that the brain is not associated with sensation or with thought. Aristotle regarded the brain simply as an
agent for cooling the heart, and preventing it from being over-
heated.   This cooling process, he considered, was effected
by the secretion of Phlegm (*pituita*), an idea still preserved in
our anatomical term *pituitary body*.   Aristotle was, in general,
much weaker in Physiology than in Morphology. Thus he
made no proper distinction between arteries and veins,

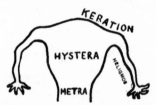

Fig. 14.—The mammalian uterus reconstructed from descriptions of
Aristotle. The terms *hystera* and *metra* persist in modern
anatomical nomenclature as does *keration* in its Latin form *cornu*.

and he believed that the arteries contained air as well as blood.
He failed, too, to trace any adequate relations between the
sense organs, the nerves, and the brain.

On the other hand, Aristotle gave fairly good descriptions
of the branches of the *vena cava*, of the superficial vessels of
the arm, of the generative and digestive organs of cephalopod
molluscs, and of many other parts of many other animals.
His general description of the vascular system is, however,
difficult. Thus he describes the heart as consisting of only three
chambers, though he perhaps has a reference—certainly the
first in history—to the *ductus arteriosus*. He realized that the
arteries are usually accompanied by veins.

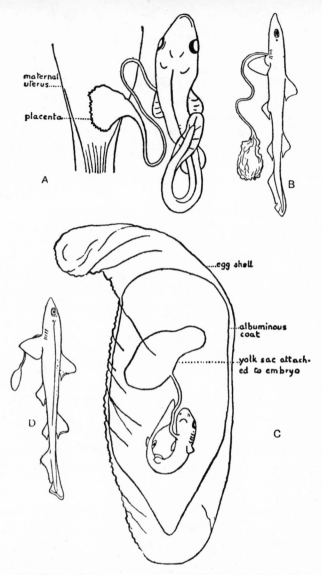

Fig. 15.—Illustrating development of placental dogfish. Redrawn from Johannes Müller, *Ueber den glatten Hai des Aristoteles*, Berlin, 1842.

    A.   Embryo of *Mustelus lævis* linked to uterus.
    B.   Embryo of *M. lævis* with placental yolk sac separated from uterus.
    C.   Egg of *M. vulgaris*.
    D.   Young free swimming form of *M. vulgaris*.

It will be seen that in the two closely allied forms, one develops in its mother's womb attached by a placenta (A, B) while the other develops external to the mother's body in an egg case (C, D).

Many of the minor works in the *Corpus Aristotelicum* are doubtless not by Aristotle himself but by later members of the *Peripatetic School*, as the followers of Aristotle were called.   Parts even of the great Aristotelian biological works are the work of these Peripatetic philosophers.   The additions supplement the genuine works of the master in some points.   Menon, a pupil of Aristotle and a physician by profession, deserves some mention in this connexion.   He prepared a *History of Medicine*.   This work is now lost, but lectures on it appear to have been given at Alexandria to a late date.   A record of these lectures has survived in the form of a papyrus fragment of a student's notebook written about A.D. 150.   It contains hints of anatomical and physiological developments and notably of actual experiments not recorded elsewhere.   Among them is a feeding experiment, which seeks to account for loss and gain of weight of a living animal.   It is the first recorded attempt to apply exact measurement in tracing known physical laws in biological phenomena.

The greatest pupil of Aristotle, Theophrastus of Eresus (370–287 B.C.), was his successor as head of his school, the Lyceum.   The chief surviving scientific treatises of Theophrastus deal with plants.   We have, however, fragments of certain other works by him.   These show that he dealt also with animal topics and departed from his master's view that the heart is the seat of the intellect.   We have a large portion of an important work by Theophrastus, *On the senses*, that is of a psychological character, but has important physiological implications.

It is appropriate to mention here the Hippocratic treatise *On fractures and dislocations*.   The main part of this work is not later than about 350 B.C., and it was known to Diocles. It contains clear and accurate, but very elementary, anatomical ideas.   There are, however, anatomical passages in it which are much more advanced and appear to be of considerably later date.   These interpolated passages may either represent the standard of Athenian anatomy in its latest stage—say about 290 B.C.—or they may be of early Alexandrian origin. One such paragraph describes the structures around the shoulder-joint.   This is peculiarly interesting and important

PLATE IV

METROLOGICAL RELIEF OF FIFTH CENTURY B.C.

Recording conquest of Samos by the Athenians under Pericles in 440.
It gives the measure of the Fathom and the Foot. Original in
Ashmolean Museum at Oxford.

HEAD OF A PHILOSOPHER, ONCE THOUGHT
TO BE ARISTOTLE

Bronze in Naples Museum from Herculaneum. Work of
First Century A.D. from an earlier original.

[face p. 22

as the first clear description of the surgical application of knowledge admittedly derived from dissection of the human body. The dating of the work and of the parts thereof therefore deserve more intensive investigation than they have yet received.

We should here refer also to a series of fanciful and corrupt anatomical treatises which probably date from the very end of the Athenian period, perhaps about 300 B.C. Among the minor works of the Hippocratic Collection are those *On the Nature of bones, On fleshes, On glands,* and *On the Humours.* The titles bear but little relation to their contents. Thus that *On the bones* deals chiefly with an imaginary scheme of distribution of the veins, that *On fleshes* is a confused and difficult work describing the development of the fœtus and endeavouring thereby to support the philosophy of Heracleitus (see p. 10). These treatises contain little positive observation, and had no influence on the course of anatomical history. Their character shows why the future of Anatomy did not lie in Athens. We can afford to pass them by.

§ 5 *Aristotelian Philosophy in its bearing on anatomical thought*

Before we pass to the events of anatomical history in the period that succeeded the death of Aristotle, it is reasonable to pass in review the elements in the Aristotelian philosophy which affected later anatomical thought. For more than two thousand years Aristotelian philosophy, in more or less corrupted form, constituted the main intellectual pabulum of mankind. Without, therefore, some knowledge of the biological verdicts of Aristotle it is impossible to understand the subsequent history of anatomical thought.

The problem of the nature of generation is one in which Aristotle never ceased to take an interest. Among the methods by which he sought to solve it was embryological investigation. In his ideas on the methods of reproduction we must seek also the main bases of such classification of animals as he exhibits. His most important embryological researches were made upon the chick. He asserts that the first signs of development are noticeable on the third day, the heart being visible as a palpitating blood-spot. As it develops two meandering blood-vessels extend to the surrounding tunics. A little later he observed that the body had become distinguishable, and was at first very small and white, the head being clearly distinguished and the eyes very large. To follow the main features of the later stages was a comparatively easy task.

Aristotle was greatly impressed by these phenomena. He lays great stress on the early appearance of the heart in the embryo. Corresponding to the general gradational view that he had formed of Nature, he held that the most primitive and fundamentally important organs make their appearance before the others. Among the organs all give place to the heart, which he considered the first to live and the last to die. There, as we have seen, he placed the seat of the intelligence.

Thus, not only in his account of the "Ladder of Nature" (Fig. 12) but also in his theories of individual development, Aristotle exhibits some approach to evolutionary doctrine. This is somewhat obscured, however, by his peculiar vein of the nature of procreation. On this topic his general conclusion is that the material substance of the embryo is contributed by the female, but that this is mere passive formable material, almost as though it were the soil in which the embryo grows. The male, by giving the principle of life, the *soul (psyche)*, contributes the essential generative agency. But this *soul* is not material, and it is, therefore, not theoretically necessary for anything material to pass from male to female. The material which does in fact pass with the semen of the male is, as the older philosophers would have said, an *accident*, not an *essential*. The essential contribution of the male is not matter but *form* and *principle*.

The female then only provides the *material,* the male the *soul*, the form, the principle, that which makes life. Aristotle was thus prepared to accept instances of fertilization without material contact, i.e., in effect, parthenogenesis. In the centuries that came after him such instances were not infrequently adduced, and this doctrine was given a special turn by Christian theologians. Belief in the "accidental" character of the material contribution of the male was common among men of science till the nineteenth century. The general attitude as to the nature of fertilization as set forth, for instance, by Harvey in his book, *On the generation of animals*, published in London in 1651, is practically identical with the views of Aristotle just 2,000 years earlier.

We must say something concerning Aristotle's conceptions of the nature of Life itself. He was before all things a " vitalist ". For him the distinction between living and not-living substance is to be sought not in its material constitution, but in the presence or absence of something that he calls *psyche*, which we translate *soul*. His teaching on this topic had the profoundest influence on subsequent anatomical and physiological thought. That teaching is to be found in his great book *On soul* (*De anima*). He does not there regard matter as organic or inorganic—that is a distinction of the seventeenth century physiologists—nor does he think of things as animal, vegetable, or mineral—that is a distinction of the mediæval alchemists—but he thinks of things as either *with soul* or *without soul* (*empsycha* or *apsycha*).

Aristotle's theory as to the relation of this *soul* to material things is a difficult and complicated subject. Its adequate discussion would take us far beyond our theme. He holds, however, that the soul is related to the idea of *form*. Matter is for him identical with potentiality, form with actuality. In living things, then, the soul is that which gives the form or actuality. He defines life existing in matter as " the power of self-nourishment and of independent growth and decay ". Of the soul, the principle of life, he distinguishes three orders or types, the *vegetative* or nutritive and reproductive, the *animal* or sensitive, and the *rational* or intellectual soul. The last he at first held was peculiar to man, but later he modified this view.

Aristotle does not make any formal classification of animals. Scattered through his works are many terms employed in a way which suggests that they might be developed for classificatory purposes. By examining his definitions of these terms we are enabled to draw up this arrangement of animal forms which we may reasonably regard as the Aristotelian classificatory scheme :—

ENAIMA (*Sanguineous and either viviparous or oviparous*) = *vertebrates*.

Viviparous in the internal sense.

1. Man.
2. Cetaceans.
3. Viviparous quadrupeds.
   (a) Ruminants with incisor in lower jaw only, and with cloven hoofs.
   (b) Solid-hoofed animals.
       (i) Equidae.
       (ii) Other solid-hoofed animals.
   (c) Other viviparous quadrupeds.

Oviparous though sometimes *externally* viviparous.

With perfect ovum.

4. Birds—
   (a) Birds of prey with talons.
   (b) Swimmers with webbed feet.
   (c) Pigeons, doves, etc.
   (d) Swifts, martins, etc.
   (e) Other birds.
5. Oviparous quadrupeds = Amphibians and most reptiles.
6. Serpents.

With imperfect ovum.

7. Fishes—
   (a) Selachians = Cartilaginous fishes and, doubtfully, the "fishing frog" (*Lophius piscatorius*).
   (b) Other fishes.

ANAIMA (*Non-sanguineous and either viviparous, vermiparous, or budding*) = *invertebrates*.

With perfect ovum.
With "scolex".
With generative slime, buds or spontaneous generation.
With spontaneous generation only.

8. Cephalopods.
9. Crustaceans.
10. Insects, spiders, scorpions, etc.
11. Molluscs (except Cephalopods), Echinoderms, etc.
12. Sponges, Coelenterates, etc.

Some of the elements in this classification are fundamentally unsatisfactory in that they are based on negative characters. Such is the group of *Anaima* which is paralleled by our own equally convenient and negative though morphologically meaningless equivalent *Invertebrata*. Others, such as the subdivisions of the viviparous quadrupeds, can only be somewhat forcibly extracted out of Aristotle's text. But there are yet others, such as the separation of the cartilaginous from the bony fishes, that exhibit true genius and betray a knowledge that can only have been reached by careful investigation. Remarkably brilliant too, is his treatment of Molluscs.

We note that the modern terms *species* and *genus* are Latin translations of terms that Aristotle employs. The *species* of the Aristotelian works are substantially the same as ours, but the Aristotelian employment of the word *genus* is much looser.

Among the most enduring of all the Aristotelian conceptions were not his finely thought out biological theories, but a doctrine of the constitution of matter, of which the modern student hears nothing (Fig. 16). He held, following more ancient writers, that there were four primary and opposite fundamental qualities, the *hot* and the *cold* ; the *wet* and the

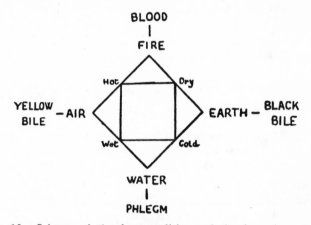

FIG. 16.—Scheme of the four qualities and the four elements as exhibited in the Aristotelian writings and of the four humours as exhibited in the Hippocratic and Aristotelian writings.

dry. These met in binary combination to constitute the four essences or existences which enter in varying proportions into the constitution of all matter. The four essences, or to give them their usual name, *elements*, were *earth, air, fire,* and *water* (Fig. 16). Thus water was wet and cold, fire hot and dry, and so forth. With this theory, later writers combined the Hippocratic doctrine of the four humours—*blood, phlegm, black bile* (melancholy), and *yellow bile* (choler). The idea, now departed altogether from our scientific discipline, still persists embedded in our language. In the authorized version of the Bible, St. Paul speaks of being " in bondage under the

elements " which he calls " weak and beggarly " (Galatians
iv, 3 and 9), and again, St. Peter, describing the end of the
world, pictures that " the elements shall melt with fervent
heat ".. Poetry still uses such ideas as the " raging of the
elements " and " elemental forces ". We may yet speak
of a " fiery nature " or an " aerial spirit ", while terms derived
from the doctrine of the four humours remain to this day in
popular, if not in scientific, medicine. Thus we speak even
now of a *sanguine*, a *phlegmatic*, a *melancholy*, or a *choleric*
disposition, and such words conjure up real pictures in our
minds. Until it began to be undermined by Robert Boyle
(1627–1691), in the seventeenth century, the doctrine of the
four elements persisted in its entirety, while ideas and terms
derived from the old humoral pathology can, in fact, be traced
in the medicine of the twentieth century.

§ 6   *The   Great   Alexandrians,   about*   300   B.C.–250   B.C.

(*a*) *Herophilus, the Father of Anatomy*

The Athenian scientific school came to an end in the
generation after the loss of Athenian liberty. The important
terminal events are the death of Alexander in 323, with the
consequent break up of the Alexandrian Empire, the death
of Aristotle in 322, and the death of Theophrastus in 287.

From now on for several centuries the centre of the
scientific world was the city of Alexandria in Egypt, where the
Ptolemaic kings established a library and museum. It was at
Alexandria that Anatomy first became a recognized discipline.
The two earliest and greatest Alexandrian anatomical teachers
of whom we hear are Herophilus and Erasistratus, who were
both Asiatics and who both flourished in the first half of the
third century B.C. They inaugurated what we may call the
" Alexandrian period " of Anatomy. The works of both these
men are lost, but we gain a good idea of them from passages
gleaned from Galen.

Herophilus of Chalcedon (about 300 B.C.) was, so Galen
assures us, *the first to dissect both human and animal bodies*.
This remark must surely refer to *public* dissection, since, as
we have seen, we have considerable evidence of dissection,

PLATE V

PTOLEMEY I SOTER (reigned 306-283 B.C.)
Founder of the Alexandrian Medical School.
Head in Louvre from Greece. Work of early
Third Century, B.C.

HIPPOCRATES (c. 460-c. 370 B.C.)
Bust in British Museum. Work of Second or Third
Century, B.C.

and even of human dissection, at an earlier date, notably in the case of Diocles. Herophilus was a very successful teacher, and wrote a work *On anatomy*, a special treatise *Of the eyes*, and a popular handbook for midwives. Despite the loss of his works, we know that the anatomical achievements of Herophilus were very numerous. He definitely recognized the brain as the central organ of the nervous system, and he regarded it as the seat of intelligence, thus reversing the views of Aristotle on the primacy of the heart. He was the first to grasp the nature of nerves other than those of the special senses. He divided nerves into motor and sensory, but continued to use the word *neuron* (Latin, *nervus*) for sinews and ligaments. He described the meninges and the *torcular Herophili*, which is named after him. The term *rete mirabile* is a Latin translation of the title which he gave to that structure, and the fact that he described it shows that he worked on animals, since it does not exist in man. He greatly extended the knowledge of other parts of the brain, distinguishing the cerebrum and cerebellum, the fourth ventricle, and, above all, the *calamus scriptorius*, which he named. The terms *prostrate* and *duodenum* are derived from those which he used. We owe to him also the first description of the lacteals. The observations of Herophilus on the lacteals were extended by Erasistratus (see p. 31), but were not subsequently improved upon until the publication of the work of Gasparo Aselli (1581–1626, see p. 160) nearly two thousand years later.

Herophilus made the first clear distinction between arteries and veins. Pulsation he regarded as an active process in the arteries themselves. He extended the study of the pulse. In his book for midwives, he gave an elementary account of the anatomy of the uterus. In this connexion we may observe that he is the first medical teacher recorded to have had a woman pupil, one Agnodice. The story, it must be admitted, is of more than doubtful authenticity. There has survived a considerable fragment of an anatomical treatise by Herophilus, containing a section describing the liver. His description of this organ shows that, at times at least, he did really work on the human subject.

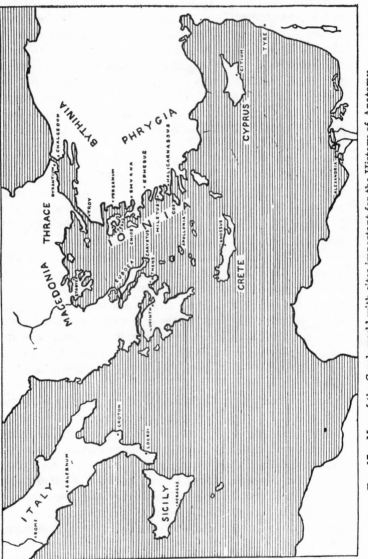

Fig. 17.—Map of the Greek world with sites important for the History of Anatomy.

(b) Erasistratus, the Father of Physiology

Erasistratus of Chios (about 290 B.C.), the rival and younger contemporary of Herophilus, was a physiologist rather than a pure anatomist. He may be said to have founded Physiology as a formal discipline in the same way as Herophilus founded Anatomy. Erasistratus was essentially a " rationalist ", and professed himself a foe to all mysticism. In the last resort, however, he had to invoke the idea of Nature as a great artist acting as an *external* power shaping the ends to which the body acts. This is in contrast with Aristotle's view of " soul " as an *innate* and not external nor even merely an internal force. The reader may be reminded that this is a statement of a problem still canvassed among biologists. Erasistratus accepted the *atomism* and the consequent " materialism " of the earlier philosopher Democritus (about 470–380 B.C.). He combined it, however, as we shall see, with a very definite pneumatic theory.

The physiology of Erasistratus was based on the observation that every organ is equipped with a threefold system of " vessels ", vein, artery, and nerve. He observed that these divide to the very limits of vision, and he considered that the process of division goes on beyond those limits. The minute divisions of these vessels, plaited together, make up the tissues. Veins, arteries, and nerves even are made of minute tubes of the same nature as themselves, through which they are nourished. Blood and two kinds of pneuma are the essential sources of nourishment and movement. The blood is carried by veins. Air, on the other hand, is taken in by the lungs and passes to the heart, where it becomes changed into a peculiar pneuma, the *vital spirit*, which is sent to the various parts of the body by the arteries. This spirit is carried to the brain, where it is further changed, apparently in the ventricles, to a second kind of pneuma, the *animal spirit*. The animal spirit is conveyed to different parts of the body by the nerves, which are hollow. The physiological system of Erasistratus was further developed by Galen (p. 58), although the " Prince of Physicians " professed great opposition to the views of his forerunner.

In the brain Erasistratus, like Herophilus, distinguished between cerebrum and cerebellum. He gave a detailed description of the cerebral ventricles and of the meninges. He particularly observed the convolutions and noted that they were more elaborate in man than in animals, and he associated this complexity with the higher intelligence of man. He considered that the cerebral ventricles were filled with *animal spirit*. He traced the nerves towards the brain, and at first regarded the *dura* as their effective termination. This conclusion was based on animal experiments, which seemed to prove that cutting the *dura* gave rise to movements. Later, as a result of further experiments, he altered his view, and traced the nerves into brain substance. He distinguished between the categories of nerves, separating sensory from motor. Nerves he regarded as conveying *animal spirit* from the brain through their supposed lumina, though later he perceived difficulties in this theory. He attained to a clear view of the action of muscles in producing movement. He regarded the shortening of muscles as due to their distension by *animal spirit*. We here note that similar theories as to the nature of muscular action were again set forth, on theoretical grounds, in the seventeenth century by Descartes (1596–1650) and Borelli (1608–79), but were rebutted by the experiments of Swammerdam (1637–80). In this connexion we may remind ourselves that we are still in the dark as to the mechanism of contraction of muscle fibre.

Erasistratus anatomized newly born goats and saw and described lacteals more clearly than Herophilus. His special investigations of heart and blood vessels led to considerable advances, so that he came very near discovering the circulation of the blood. He regarded the heart as the source of both arteries and veins, a matter in which he was not only ahead of his time, but was ahead of all opinion until Harvey.

Erasistratus perceived that, although arteries were empty in dead bodies, yet when incised during life they manifestly contain blood. He explained this as due to escape of pneuma through the wound leading to a vacuum. As a result of the formation of this vacuum he considered that *blood was sucked into the arteries from the veins through very fine inter-*

communications between the two types of vessel. In other words, he realized the existence of the capillary system. The view that the arteries contained air was disproved by an experiment of Galen, about four hundred and fifty years later, but the fact that Erasistratus realized that communication existed between the veins and arteries is very remarkable and greatly to his credit.

Erasistratus regarded the right ventricle of the heart as filled with blood and the left ventricle with vital spirit. During diastole blood was drawn into the right ventricle and pneuma into the left ventricle, and these were expelled during systole. The return of blood and vital spirit was prevented, so Erasistratus considered, by the semi-lunar valves. He hit on the function of the tricuspid valve, which owes its name to him. The office of the bicuspid, he considered, was to prevent vital spirit from leaving the heart save by the aorta, in the same way as the tricuspid prevented regurgitation of the blood. It will thus be seen that he was only hindered from reaching the idea of the circulation of the blood by his pneumatic theory. The auricles were regarded by him as part of the pulmonary vessels. He knew and described the following vessels, among others, Aorta, Aorta descendens, Pulmonary artery, Intercostal arteries, Hepatic arteries, Renal arteries, Gastric arteries, Pulmonary veins, Vena cava, Azygos vein, and Hepatic veins.

§ 7 *Decline of the Alexandrian School, about* 250–50 B.C.

Anatomical research at Alexandria flagged after the first generation, but the city long remained a great teaching centre, and minor advances were made. The stagnation in medical matters at Alexandria is in sharp contrast to the continued activity there in Mathematics, Astronomy, Mechanics, and Geography.

Of the later Alexandrian medical writers we know little. Of one, Hegetor, however, who lived about 130 B.C., we have a fragment from which we learn that he knew of the *ligamentum teres* of the hip joint. Another was Apollonius of Citium, an Empiric physician who studied at Alexandria in the first century B.C.   He wrote a commentary on the Hippocratic work *Fractures and Dislocations*.   There is in the library of Lorenzo de Medici at Florence an illustrated manuscript of the ninth century A.D. containing this commentary.   The manuscript is copied from an exceedingly ancient original, perhaps even contemporary with Apollonius himself.   The figures in this most interesting document must have exhibited considerable anatomical knowledge, as can be seen even from the inferior drawings of the copy.   (Plate VI.)   The Florence manuscript was studied in the sixteenth century by the artist Primaticcio (1490–1570), and by the anatomist Vidus Vidius (died 1569, see p. 144), the friend of Benvenuto Cellini (1500–71).   The figures of Vidius reproduce those of this codex, and were influential in the development of surgical procedure in the sixteenth century.

§ 8 *Human Vivisection at Alexandria*

The names of Herophilus and Erasistratus are linked with the charge of dissecting living men.   The evidence rests on Celsus (about 30 B.C.) and Tertullian (about A.D. 155–222). The charge was repeated by St. Augustine (A.D. 354–430), at a much later date.   Tertullian was violently anti-pagan and accuses Herophilus of being a butcher who dissected six hundred (living ?) persons.   In another passage in which he reprobates embryotomy he accuses Herophilus of the death not only of fœtuses, but of adults.   The *odium theologicum* which colours the work of Tertullian cannot be ascribed to Celsus, who was a pre-Christian writer (see p. 39) and who twice refers to the charge of vivisection.   Celsus utterly reprobates such practice, though he is in favour of the dissection of the dead.

Medical critics have occupied themselves in answering this charge.   Their chief rebutting arguments are :—

(*a*) Neither Tertullian nor Celsus were medical men.

(*b*) The practice of dissection at Alexandria, being an innovation, was likely to excite adverse criticism.

PLATE VI

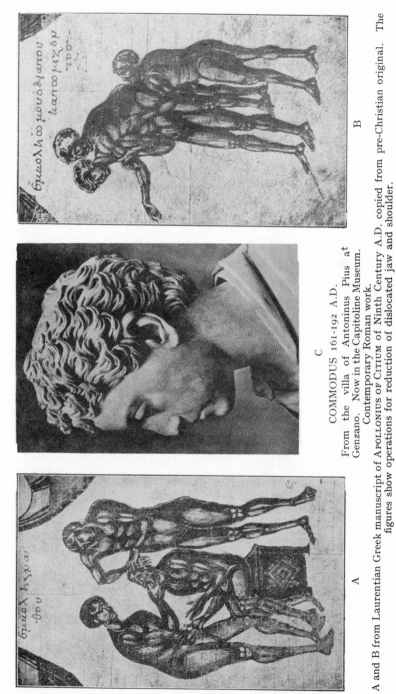

A

C

B

COMMODUS 161-192 A.D.
From the villa of Antoninus Pius at
Genzano. Now in the Capitoline Museum.
Contemporary Roman work.

A and B from Laurentian Greek manuscript of APOLLONIUS OF CITIUM of Ninth Century A.D. copied from pre-Christian original. The
figures show operations for reduction of dislocated jaw and shoulder.

[face p. 34

(c) Prejudice against dissection exists always and everywhere, and even in modern times excites rumours and calumnies. The charge of human vivisection has often been falsely made in modern times, e.g. against Berengar of Carpi, Vesalius, and Fallopius.

(d) All later physicians, including the garrulous Galen, are silent on the point.

The last argument is much more convincing than might, perhaps, at first be thought. Galen had to rely on animals for his anatomical investigations. He habitually experimented on living animals, and records his results and experiences in many places. He wrote a work on the technique of physiological investigation. That work is now lost, but Galen very often repeats himself and we have very numerous references to vivisection experiments in his other writings. Galen was extremely antagonistic to the views of Erasistratus and his followers, and devotes two books, which still exist, to their denunciation. If therefore Galen *disapproved* of human vivisection, he would certainly have cast it up against Erasistratus, for he was far from adverse to violent criticism of those from whom he differed. Had he *approved* of human vivisection—as he did of human dissection—he would surely have referred to the practice by the Alexandrian anatomists had he regarded the rumours as more than mere vulgar reports. The complete silence of Galen through the hundred and twenty-seven separate works ascribed to him is thus a very impressive rebutting argument.

§ 9   *The Alexandrian Anatomists and the Wisdom Literature*

During the third century B.C. Alexandria became an important Jewish centre. Parts of the Old Testament were rendered into Greek about 250 B.C. The contact between Hellenic and Hebraic culture had very important effects on the Hebrew view of Nature. Thus, while the earlier Biblical literature contains many references to divine intervention, the later so-called " Wisdom Literature " practically ignores the supernatural in physical matters. Hellenism in this literature has affected the Hebrew views as to the physical constitution of the world, and this effect occasionally passes into the domain of Anatomy. The seat of the understanding in the Wisdom Literature is usually placed in the heart (e.g. *Ecclesiasticus* xvii, 5–6 and 16–18), following the Aristotelian tradition and contrary to the view of Herophilus and Erasistratus ; contrary too, to an earlier Hebrew view which places it in the liver (*Psalms* xvi, 7, *Proverbs* xxiii, 16, and again later in Revelation ii, 23). In several places in this Wisdom Literature there are also traces of the doctrine of the four elements (e.g. *Wisdom of Solomon* vii, 17, and xix, 18–21). Again, in Aristotle's *Generation of animals* there is a theory that the substance of the embryo is formed out of the catamenia which are not discharged during pregnancy. This view was adopted by the Alexandrian school and reappears in the *Wisdom of Solomon* (vii, 2). In *Ecclesiasticus* (xlii, 24, and xxiii, 13–15), too, we encounter the " doctrine of opposites ", as set forth by Empedocles and popularized by Aristotle.

With the absorption of Egypt into the Roman Empire, and with the final extinction of the Ptolemaic dynasty by the death of Cleopatra in 30 B.C., Alexandria ceased to have great scientific importance. The Alexandrian school continued for centuries with restricted activity and devoid of all originality. Intellectually, however, it became subordinate to the Metropolis. The future of Anatomy must be considered from the point of view of the Roman Empire.

II
THE EMPIRE AND THE DARK AGES
50 B.C.–A.D. 1050.

§ 1 *The Beginnings of Anatomy at Rome,* 50 B.C.–A.D. 50

WITH the advent of the Empire and of Imperial ideas, the tone of Science changes. There is a waning in enthusiasm for all save " useful " knowledge. The so-called " practical " outlook has been and is a great obstacle to the advance of Medicine. The plaintive cry for what is " useful " as against " theoretical knowledge " echoes down the ages. We hear it still. But when and where that cry swells into a chorus, then and there Science dies. So it was of old in Rome. Greek curiosity as to the causes of things had long been fading. Thus the activities of those interested in phenomena became at last devoted to the task of compilation from approved authors. During the century and a half between 200 B.C. and 50 B.C. Rome had been steadily fastening her hold on the Eastern Mediterranean, and there was a corresponding depression in the activity of the Alexandrian and other Eastern Schools.

In Rome itself a medical school was founded about 60 B.C. by Asclepiades of Bythinia. Asclepiades introduced the atomic view of Democritus into Medicine, but, despising Anatomy, he is unimportant for our purpose. The medical school at Rome was at first a mere personal following of the physicians, who took pupils and apprentices with them on their visits. Toward the end of the reign of Augustus (died A.D. 14), such groups combined to form colleges, which obtained their own meeting place, the *Schola medicorum*. The school became further organized and enlarged under Vespasian (A.D. 70–9), and teachers were given a salary at the public expense. The system was extended and enlarged under Hadrian (A.D. 117–38), and Alexander Severus (A.D. 222–35). It probably came to an end with the death of Theodoric in 526. In its most active period minor schools were established at Marseilles, Bordeaux, Saragossa, and other provincial towns.

Dissection of the human body was still occasionally practised at Alexandria towards the end of the first century B.C., but it had ceased by the middle of the second century A.D. There is evidence that it was not then practised elsewhere. Considering the indifference to human life which the Romans exhibited, considering their brutality to slaves and the opportunities afforded by gladiatorial combats, considering the value of Anatomy for surgical practice and the demand for surgical skill involved in the organization of medical service throughout the Empire, it is truly remarkable that the anatomical knowledge of antiquity was allowed to lapse. Its passing is one of the innumerable instances that illustrate the danger of entrusting things of the mind to the tender mercies of the " practical man ". Anatomy did not revive till the rise of the mediæval universities.

§ 2   *Latin Anatomical Literature*

Medical instruction and professional intercourse in Rome and throughout the Empire continued to be in Greek, which was understood by all educated men. Latin came only very slowly into general use for medical purposes, and by the time it had been accepted as a medium for this purpose Medicine had fallen into complete decrepitude. We have, however, two non-professional Latin works which throw light on the spread of Greek anatomical and biological conceptions in Roman Society in the last pre-Christian century. These are the *De natura deorum* of Cicero (about 77 B.C.), and the *De rerum natura* of Lucretius (about 60 B.C.). In the work to which the name of Celsus is attached, written about 30 B.C., we have a definitely scientific section on Anatomy, but it, too, is not by a medical man, nor, indeed, by a practical dissector.

Cicero (106–43 B.C.) was considerably interested in medical matters. In the *De natura deorum* he discusses the nature and origin of the Universe, and in the course of this discussion he gives an elementary popular exposition of Anatomy and Physiology. The account contains the first formal statement of that teleological view of the human body which was to find fullest expression at the hand of Galen. We shall defer our consideration of that view until we discuss its greatest medical exponent.

The *De rerum natura* of Lucretius is a poem describing the physical constitution of the Universe. Lucretius was a follower of Epicurus (342–270 B.C.) and of Democritus (about 410 B.C.). His theory is *atomic,* and he believed that nothing exists but atoms and " the void " (*inane*). Everything springs from " determinate units " (*semina certa*). The genesis of all things is typified by the generation of organic beings, and the species of plants and animals are models for all processes and natural laws. The atoms, of which all things are composed, being uncreated are also indestructible. Lucretius sets forth a theory of the origin of animals and men which distantly recalls Darwin's doctrine of " Natural selection ". In spite of the enthusiasm that Lucretius exhibits for his theme, however, it is surprising how little real evidence there is to be found in his writing for any close observation of phenomena. His extreme determinism is of interest in connexion with the subsequent development of anatomical thought and notably in relation to Galen. As an observer he is negligible.

Very different in type from either of these two Latin works is that to which the name of Celsus is attached. The *De re medica* of Celsus is the earliest and best Latin medical work (about 30 B.C.). He must not be confused with the second century opponent of Christianity of the same name. Concerning Celsus, the medical writer, we know almost nothing. He was author of a large encyclopædia, of which the *De re medica* is the only complete surviving section. There are, however, fragments of the section *On agriculture* embedded in the work of Columella (about A.D. 80). It is highly probable that Celsus was not a medical man, and almost certain that his treatise is adapted from a lost Greek original. Many passages in it can be traced to the Hippocratic Collection. The work is in eight books, of which the last two deal with *Surgery.* It is our best representative of Alexandrian surgical

practice, and, except perhaps the *Commentary* of Apollonius of
Citium, it is also the earliest. The surgical section of Celsus
contains a certain amount of anatomical description which
includes a complete account of the skeleton. Among the best
anatomical descriptions to be found in it are the accounts
of the eye, of the humerus, of the radius and ulna, of the tibia
and fibula, and of the tarsus.

Celsus writes excellent and easy Latin and is perhaps the
most readable and best arranged ancient medical author. It is
thus the more remarkable that his work was unknown for
centuries and was not rediscovered till the Renaissance, when,
however, it had very considerable influence (p. 104). The *De re
medica* is the only important medical work written in Latin
during Imperial times. The remaining serious medical authors
of the period all used the Greek tongue.

We ought not to leave the Latin writers of the Empire
without saying a few words concerning Pliny (A.D. 23–79).
That voluminous, industrious, unphilosophical, gullible,
unsystematic old gossip was more widely read during the
Dark and Middle Ages than any other writer on natural knowledge. Scientifically, he is himself quite worthless, though the
superstitions that he records are a rich mine for the folklorist
and are themselves the sources of many vulgar beliefs current
to this day. The popularity of his huge work partly reflects
and partly explains the ignorance of Anatomy in the thousand
years that led up to the thirteenth century. Pliny tells
us that in his time the examination of the human viscera
was looked upon as impious. He gives as reason the
disgusting practices of the Greeks, whom he loathed and
despised. He accuses Greek medical writers, for instance, of
having enlarged upon the distinctive flavours of each of the
organs of the human body! His charge is unsupported, so
far as I know, by any serious external evidence, and on the
unsupported statement of Pliny no dog should be hanged.
We shall not delay by attempting to glean anatomical mis-
information from his work.

It is appropriate to refer here to the ancient votive offerings
of anatomical import. From the earliest times until the
present day, it has been the custom to place in temples,

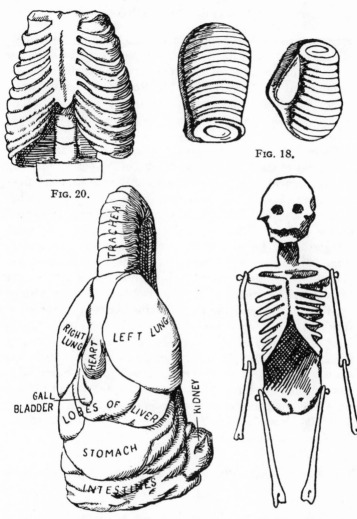

Fig. 18.

Fig. 20.

Fig. 21.

Fig. 19.

SOME OBJECTS ILLUSTRATING POPULAR ANATOMY UNDER THE ROMAN EMPIRE.

Fig. 18.—Terra-cotta models of the uterus found on " Island of Aesculapius " (now Isle of San Bartolomeo) in the Tiber. They are now in the National Museum at Rome. One perhaps shows the ovary lying by the side of the uterus.

Fig. 19.—Silver *larva* or articulated skeleton somewhat restored. It is now in the Antiquarium at Dresden.

Fig. 20.—Small marble model of the thorax, used either as a votive or perhaps, like Fig. 19, as a *memento mori*. It is of Roman origin and is now in the Vatican Museum at Rome.

Fig. 21.—Clay model of the viscera of an animal perhaps used as a votive by a *haruspex*. The identifications of the organs are written on it. It is now in the *Etruscan Museum* at Florence.

churches, and other holy places, models of a diseased or painful part. Such offerings to the deity may be expected to keep him in mind of the sufferings of the donor. The custom was particularly rife in Roman Imperial times. Many votives of Roman origin can hardly be distinguished from those which are to be seen to-day in some continental churches.

These Roman models give an idea of the anatomical knowledge of the " man in the street " of the period. Among the commoner are terra-cotta figures of the uterus, sometimes exhibiting a body lying by the side, which has been interpreted as the ovary (Fig. 18). These, it is believed, were dedicated to the god by childless women. In somewhat the same category may be placed the little images of the skeleton (Fig. 19), made of precious metal which were sometimes placed on the table or brought in with the wine at banquets to remind the guests of their mortality. There is also in existence a little marble model of the skeleton of the chest (Fig. 20), which may have been used in the same manner. Most remarkable of all are models of the complete viscera (Fig. 21), which were perhaps used as votive offerings by the *haruspices*, or " entrail observers ", a class of soothsayer in Rome whose art consisted in ascertaining the will of the god from the appearance of the viscera of sacrificial victims. This last group is thus in some ways comparable to the Babylonian liver models (p. 8).

§ 3   *Greek Anatomical Writers of the Early Empire, about* A.D. 50–150

Among the more interesting of the anatomical writers of the Empire was Rufus of Ephesus, who studied at Alexandria about A.D. 50. He produced a great number of works, some of which still survive. These were known to the Arabic writers, but did not directly influence the Middle Ages in the West. Rufus was not printed until after the middle of the sixteenth century (1554). His work *On the naming of the parts of the body* is the first devoted to the important subject of anatomical nomenclature. The names that we use nowadays for many of the structures in the eye are here encountered (Fig. 22). We observe that he describes the real form of the crystalline lens, the structure, function and position of which were misinterpreted by Galen and were almost universally misunderstood until the seventeenth century.

The *Anatomy of the bodily parts* of Rufus is a short work, which is in part a repetition of that on nomenclature. The ascription to Rufus is doubtful. It contains the first statement

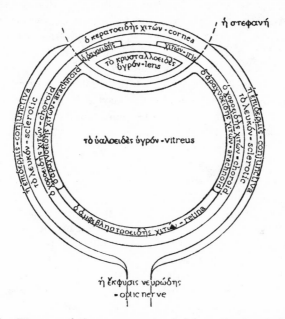

Fig. 22.—Diagram of the eye reconstructed from the descriptions of Rufus of Ephesus (about A.D. 50). The Greek names of the parts are given, together with the modern technical titles.

that the liver is five-lobed, an idea taken from the anatomy of the dog, which runs through the history of Anatomy to the sixteenth century, and was held even by Vesalius in his earlier years. The description of the heart accords in the main with that of Erasistratus, but admits that blood is a normal content of the arteries while claiming that they also contain some pneuma. Muscle, the writer tells us, following Erasistratus, is made up of arteries, veins, and nerves, is without sensation, and is the organ of movement. The description of the nerves is based on Herophilus and Erasistratus, and shows that the work was largely copied. Rufus *On bones* is another short work of doubtful authenticity, but is inferior to those of Rufus that we have already discussed.

The *Synopsis on the pulse* of Rufus is a brave attempt to base the whole of Pathology on Anatomy and Physiology, the first of its kind that has come down to us. Similar but less complete essays had been made by Herophilus and Erasistratus. Some passages in this work of Rufus claim that it is in systole that the apex of the heart strikes the chest wall. This observation is of great importance, and it is regrettable that the work was not better known in after ages. For centuries a common and natural fallacy concerning the pulse and heart-beat was that the heart struck the chest during diastole, and that therefore the pulse was synchronous with the expansion of the heart. Harvey in the seventeenth century had to perform many experiments before he could convince himself that this was false. In his hands this discovery was one of the first steps that led to the inference of the circulation of the blood.

Soranus of Ephesus is a writer of whose personality we know next to nothing. He was educated in Alexandria and seems to have practised in Rome about A.D. 100. His work *On diseases of women* was only discovered in the nineteenth century though fragments of it are preserved in the fourth century Greek writer Oribasius (325–403), and in a sixth century Latin abstract circulated in the Middle Ages under the name *Muscio*. Early MSS. of this Latin version have come down to us. They are of interest as preserving one of our few and faint memories of anatomical illustration in antiquity (Fig. 23). It is also said that Soranus prepared a diagram showing the complete anatomy of a pregnant woman. Certain mediæval diagrams may be derived with many removes from this very figure. The text of Soranus was without direct influence on the course of Anatomy, as his work was not discovered till the nineteenth century (1838). The Latin abstract of *Muscio* was, however, very widely read in the Middle Ages.

Soranus was a voluminous medical author. The only work
by him that has survived, in Greek, that *On diseases of women,*
was presumably intended for midwives, who must have been
remarkably well educated at that period. It is evident that
gynæcology had been on a thoroughly scientific basis. It
is best to use the pluperfect tense as there is evidence that,
excellent though it is, it is yet borrowed work. The account of
the uterus that it contains is one of the best pieces of ancient
descriptive Anatomy.

F IG. 23.—The uterus from the Latin summary of Soranus by Muscio.
The figure is taken from a manuscript of about A.D. 850 in the
Royal Library at Brussels (MS. 3714, folio 16). The parts *fundus,*
basis grandis, cervix, collum, and *orificium* are marked.

Marinus of Tyre worked in Alexandria about the beginning
of the second Christian century, and wrote several anatomical
treatises which Galen utilized.   Galen gives him high praise,
and in doing so provides a glimpse of anatomical teaching in
his day. He says that Marinus had collected many anatomical
observations, and that everything described in his writings
he had himself touched with his own hands and seen with
his own eyes.   Thus Marinus established " bone drill " on
firm foundations and was specially diligent in seeking out the

foramina in the skull and vertebral column. Galen provides us with an outline of the anatomical work of Marinus.

Quintus was a pupil of Marinus. He probably learnt Anatomy at Pergamum and Alexandria. He practised in Rome, where Galen tells us he was the most eminent physician of his time. He was banished from Rome for a reason unknown and died at Pergamum. He published nothing but influenced deeply his contemporaries and successors. He is of importance in the history of Anatomy because of his pupils, two of whom— Numisianus and Satyrus—influenced Galen.

Numisianus taught Anatomy at Corinth, where Galen went specially to hear him in about A.D. 149. Pelops was his pupil.

Satyrus was apparently a mere echo of Quintus, whose opinions he transmitted without alteration or omission. He taught at Pergamum where Galen learnt from him in A.D. 145.

Pelops taught Anatomy at Smyrna, and Galen went there to hear him in A.D. 149 and dwelt awhile in his house. Pelops wrote *Introductiones Hippocraticæ*, now lost, in which he maintained that not only nerves but also veins and arteries arose from the brain. He is said to have translated works from Greek into Latin, but he certainly wrote in Greek.

Lycus the Macedonian was another pupil of Quintus at Pergamum. He wrote anatomical works about A.D. 165, and is frequently quoted by Galen as an older contemporary. He is quoted also by Oribasius (A.D. 325–403), by Paul of Aegina (A.D. 625–90), and by later medical writers. Galen says he was the only one of the pupils of Quintus whom he did not know personally. Galen ranks Lycus much below Satyrus and Pelops, since his works are only compiled from those of Marinus. It would seem that neither Lycus nor any of the other second century anatomists were able to use the human subject, and it is clear that Anatomy was in headlong decay. For all his greatness the only result of Galen's work was to codify the researches of antiquity for after ages.

§ 4   *Galen, " Prince of Physicians ",* about A.D. 150–200

Galen of Pergamum (A.D. 129–99, Fig. 24) was, after Hippocrates, the greatest of the ancient physicians, and one of the greatest biologists of all time. His brilliance completely hypnotized the men of the Middle Ages by whom he was dubbed the " Prince of Physicians ". His work makes the " Indian summer " of the Anatomy of antiquity. So far as Anatomy goes, Galen's spiritual ancestry may be largely gathered from what has been said concerning the later Alexandrian writers. He seems, however, to have owed no debt to Soranus, whom he never mentions.

Galen's father, of whom he speaks with great admiration, was the architect and mathematician, Nikon of Pergamum. Galen himself was born in that city in 129, and at the age of fifteen his father sent him to attend philosophical lectures. At sixteen, when it was time for him to choose a profession, his father, influenced by a dream, chose Medicine, and he began the study at Pergamum under Satyrus (see p. 46).   Even at this early period Galen was active in literary production, and wrote his work *On the anatomy of the uterus*, which is dedicated to a midwife. He continued throughout his life to be an extremely voluminous author.

When Galen was twenty he lost his father, and he then left Pergamum for Smyrna to attend the lectures of Pelops (see p. 46). While there he composed his work *On the movement of the chest and of the lung*, which reproduces the teaching of Pelops. He next went for a short time to Corinth to study under Numisianus (see p. 46), the teacher of Pelops and the pupil of Quintus (see p. 46). Soon, however, he left for Alexandria, where he completed his education. He returned to his native city, Pergamum, in 157, being now in his twenty-eighth year, and remained for four years as surgeon to the gladiators there.

In 161, at the beginning of the reign of the Stoic emperor, Marcus Aurelius (reigned A.D. 161–80, Plate VII), Galen went to Rome to seek his fortune, as many provincials were then doing. He was immediately successful in his profession, and became friendly with the consul Flavius Boëthus. At the suggestion of this patron, Galen began his two great anatomical works *On anatomical procedure* and *On the uses of the parts of the body of man.* At this period he was giving public anatomical demonstrations. About the same time Galen entered into a controversy with the aged anatomist, Martialis the Erasistratean. The surviving works of Galen that are directed against the teaching of Erasistratus are the product of this

FIG. 24.—Portrait of Galen. No bust of Galen has survived from antiquity. The only ancient representation of him is to be found in the so-called Juliana Anicia Manuscript. This magnificent illustrated codex is now in what was once the Royal Library at Vienna. It was written in the year A.D. 487 (? 512), and presented as a wedding gift to Juliana Anicia, the daughter of Anicius Olybrius, Emperor of the West in 472, and of his wife Placidia, daughter of Valentinian III. It contains a number of descriptions and paintings of herbs and a valuable text of the herbalist Dioscorides (flourished about A.D. 60). The portrait of Galen occurs on folio 3 verso and is greatly deteriorated, much of the paint being scaled off. The figure here reproduced has been prepared for this volume by Mr. T. L. Poulton, artist to the Anatomical Department at University College. Mr. Poulton has worked on enlarged photographs, has reproduced the original line by line, and has finally filled in missing details.

dispute. In response to the inquiry of Martialius as to the school of which he considered himself a member, Galen loftily replied that he followed none, but chose what was good from all, regarding it as the mark of a slave to call any man master.

In 165 plague broke out in Rome, and Galen fled to Pergamum, coming back, however, in the following year. On his return he received a command from Marcus Aurelius to join him with his assembled army at Aquileia, close to the modern city of Venice, and to accompany him thence as body physician in his expedition against the German tribes (Plate VII). The plague again broke out, and Marcus hurried back to Rome. When the plague waned, the Emperor returned to Aquileia, and proceeded thence into Pannonia (Fig. 25). Galen, however, managed to escape service with him in the field on the plea of looking after the little prince Commodus (Plate VI). He took advantage of the leisure thus afforded to complete his great anatomical works.

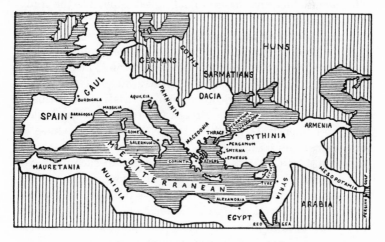

Fig. 25.—The Roman Empire in the time of Galen.

§ 5 *Galen's Anatomical Philosophy*

We must now turn to the anatomical content of the vast corpus of Galenic writings. They set forth a medical system of which the substance is based on the Hippocratic Collection and the form is derived from Aristotle. Galen's Anatomy may be examined under two aspects, which we may describe as (*a*) descriptive, and (*b*) philosophical. The philosophical aspect comes out most clearly and consistently in the *Uses of the parts of the body of man*. In that remarkable work, vastly influential in the ages which followed, Galen seeks to prove that the organs are so well constructed, and in such perfect relation to the functions to which they minister, that it is impossible to imagine anything better. Thus, following the Aristotelian principle that Nature makes nought in vain, Galen seeks to justify the form and structure of all the organs— nay, of every part of every organ—with reference to the functions for which he believes they are destined. We are thus in the presence of a work that is not, strictly speaking, a treatise either of Anatomy or of Physiology, but in which Anatomy and Physiology are subservient to a particular doctrine and are used to justify the ways of God to man. We have, in fact, the thesis of final causes applied to the study of the animal organism.

The problem of final causes is developed by Galen along definite lines. He considers that it is possible to discover the end served by every part of the animal, and, moreover, to show that such a part, being perfectly adapted to its end, could not be constructed other than as it is. To say this is to go even further than the Bridgwater treatises which under-took to demonstrate the " Power, Wisdom, and Goodness of God as manifested in the Creation ". It is to claim that in every work of Creation, and in every detail of such work, we can demonstrate these attributes along known principles. It is to claim, in fact, a complete knowledge of the Laws of Nature. No flamboyant modern man of Science, however inflated with confidence drawn from the most sweeping presentation of scientific determinism, however intoxicated with his own scientific achievements, has as yet arrogated such powers to himself. To conceive that such claims should be made by a pious theistically-minded author, the reader must think himself back into a very different philosophical environment from that to which we are nowadays accustomed.

The prevailing philosophy of Galen's world was the Stoic schemes so admirably and beautifully expounded by his royal master, Marcus Aurelius. There were, of course, other systems of philosophy in vogue, Epicurean, Gnostic, Neo-

PLATE VII

THE EMPEROR MARCUS AURELIUS (121-180)

Panel from a Triumphal Arch erected on the Capitol at Rome in 176
commemorating the Emperor's triumph over the Germans and
Sarmatians. Marcus is shown as in the field receiving two German
captives brought in by the Praetorian guard. The grave sad face
of the Emperor is very striking. The panel here represented is now
in the Palace of the Conservatori at Rome.

[face p. 50

platonic, and the rest, to say nothing of the various Oriental cults, such as that of Persian Mithra, of Egyptian Isis, of Phrygian Cybele, that were permeating the Empire. None of these systems, however, interested their followers in phenomena, nor was there any system but that of the Stoics which could make an appeal at once to men of action and to men of scientific knowledge.

Now in the World of the Stoic philosopher all things were determinate, and they were determined by forces acting wholly outside Man. The type and origin of that determination the Stoic sought in the heavens, in the majestic and overwhelming procession of the stars. The recurring phenomena of the spheres typified, foreshadowed, nay, exhibited and controlled, the cycle of man's life. Man dwelt in a finite world bounded by the firmament and limited by a flaming rampart. Within that rampart all worked by rule—and that rule was the rule of the heavenly bodies. Astrology had become one of the dogmas of the Stoic creed.

To such a world Galen's determinism was in itself no strange thought. Remember that Galen had, in his youth, been well trained in the Stoic philosophy. Yet Galen's view was far from being in accord with Stoicism. Though a determinism, it was a determinism of perfection in which all was fixed by a wise and far-seeing God, and was a reflection of His own perfection. That perfection can be traced in the body of man, and Galen exclaims outright that a knowledge of the uses of the organs reveals Deity more clearly than any sacred mysteries. Galen repeatedly adopts the argument from design for the existence of God ; indeed, it is his sole argument. Now such a scheme did not ill fit the new creed which was just beginning to raise its head and was destined to replace Stoicism and all the other pagan schemes. Galen's thought, in fact, made a special appeal to the Christian point of view, and this is doubtless the reason that a larger bulk has been preserved of his works than of those of any other pagan writer.

In several places Galen mentions both Judaism and Christianity, though without much respect. In the great anatomical work under discussion he explains that in his belief

God always works by law, and that it is just for this reason
that Natural Law reveals Him, and he adds that " in this
matter our view . . . differs from that of Moses ". It seems very
probable that he had read some books of the Bible. His
position can thus be summed up as intermediate between
Stoicism and Christianity. On the one hand he accepted the
Natural Law of the Stoic philosophy, but rejected its
astrological corollary. On the other hand he accepted the
Divine Guide and Architect of the Universe which corre-
sponded to the Christian scheme, but rejected all idea of
miracle.

Let us, however, consider the results of Galen's doctrine
of the uses of all the parts. Treated, as it must be, on the
a priori basis, it was inevitable that it should turn men away
from the observation of Nature and that it should make them
content with arbitrary solutions of the many problems which
his principle raised. In the case of Galen himself, who came as
a pioneer of this belief, it was a novel presentation of the World
which was thus still worth exploring. Galen explored it, and
his Anatomy—within certain limits—was exact.   His
teleological theory, however, removed the motive for further
exploration on the part of his successors,and,with Galen's death,
Anatomy and Physiology too fell dead and were not reborn
for a thousand years. To a brief review of that Anatomy and
Physiology we must now devote ourselves.

§ 6   Galen's Anatomical Achievement

The best presentation of Galen's actual anatomical knowledge is found in his great work *On Anatomical procedure*.
This was originally in sixteen books, of which nine only have
survived in Greek. It has long been known that there existed
an Arabic version of the remaining seven. These have recently
been published with a German translation, so that the entire
work is now accessible.

We may begin with the bones. These Galen had studied on
an actual human skeleton at Alexandria, and he describes
them in a special work *On bones for beginners*. He divides them

into long bones with a medullary canal and flat bones without such a canal. He distinguishes *apophyses*, *epiphyses*, and *diaphyses*, terms of which the first two have descended to us from him.[1] He uses the word *trochanter* in the modern sense, but it seems to have entered Anatomy through the work of his contemporary Julius Pollux (see p. 107). He had a fairly good idea of the bones of the cranium. He regarded the teeth as bones, and he gives a good description of their origin. He recognized twenty-four vertebræ terminated by the *coccyx* and *sacrum*. The latter he regarded as the *most important* bone of the spine, and the word he used to describe it was misunderstood by the Latins as equivalent to *sacred*, hence our term *sacrum*. Galen gives accurate elementary descriptions of the vertebræ, of the ribs, of the sternum, of the clavicle, and of the bones of the limbs.

In Arthrology Galen recognized two main orders of joint, to which he gives the names *diarthrosis* and *synarthrosis*. Diarthrosis, or articulation with movement, is divided into *enarthrosis*, *arthrodies*, and *ginglymus*. The synarthrosis, or articulation without movement, is divided into *suture*, *gomphosis*, and a third kind which he regards as simple linear arthrosis, as in the *symphysis pubis*. The descriptions of joints are less satisfactory than most elements in his anatomical system.

As regards the muscular system there can be little doubt that Galen's work was in large part of a really pioneer character. He wrote a special book *On the anatomy of the muscles*. Throughout his works the muscles are perhaps the structures that he described most accurately. His writings contain frequent references to the form and function of muscles of various animals. Thus the dissection of the muscles of the orbit and larynx was performed on the ox, and the muscles of the tongue are described from the ape. Occasionally, he indicates that he is aware of the difference between certain of the muscles he is describing from those of man ; instances of these are his description of the *flexor longus hallucis* and of the *lattissimo condyloideus*. A great difficulty in reading his work, however, is the absence of any properly worked out

[1] His use of the word *diaphysis* is quite different from ours.

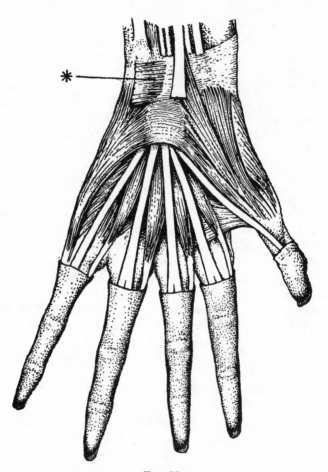

Fig. 26.

Figs. 26 and 27.—Dissection of muscles of palm of the Barbary ape, *Macacus inuus*, the animal on which Galen did most of his myological work. By its side is placed the dissection of the muscles of the human palm. It will be seen that the ape's hand contains all the muscular structures present in the human hand, though the proportional development of the various muscles differs. The difference in the two hands may be thus summarized :

(*a*) The *palmaris brevis* has a different origin. This muscle is marked by an asterisk in both hands.

(*b*) The third finger is longest in man, the fourth in the macaque.

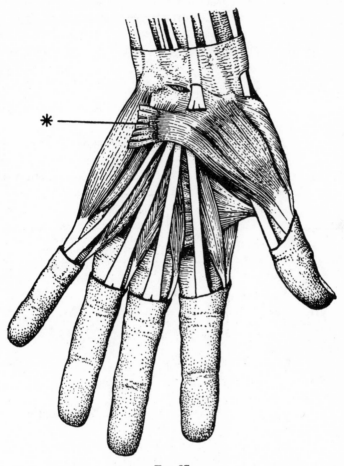

FIG. 27.

(c) The human hand is broader and relatively shorter.

(d) The human thumb is relatively much longer.

(e) The long but narrow wrist of the macaque is associated with an elongation of the muscles taking origin about the carpal ligament. This is apparent in the greater extent to which the long tendon of the thumb of the macaque is exposed.

It will be seen that the anatomical resemblance of the two species is sufficiently close for a general description of the one to be applied to the other. Galen's anatomy, drawn from the macaque, was thus a serviceable record for the crude surgical processes of the ages which followed him.

nomenclature—a defect which haunted ancient Science in general and ancient biological Science in particular. Thus he calls the *serratus magnus* " the muscle situated on the concave part of the scapula and expanding the chest " while he speaks

MODERN USAGE.	GALEN.
I. Olfactory.	Not regarded as separate nerves.
II. Optic.	" The soft nerves of the eyes."
III. Oculomotor.	" The nerves moving both eyes."
IV. Trochlear.	Not described.
V. Trigeminal.	{ " Third pair of nerves." { " Fourth pair of nerves."
VI. Abducent.	United with II.
VII. Facial. VIII. Auditory.	" Fifth pair of nerves."
IX. Glossopharyngeal. X. Vagi. XI. Spinal accessory.	" Sixth pair of nerves."
XII. Hypoglossal.	" Seventh pair of nerves."

FIG. 28.—Table of the cranial nerves according to Galen compared to modern usage.

of the *bulbo-cavernosus* simply as " the muscle at the neck of the bladder ". Anatomical nomenclature remained in this chaotic condition until the sixteenth century, when it was reformed by several authors, important among whom were

Jacques Dubois (Jacobus Sylvius, 1400–1500, see p. 108) and Adrian van den Spieghel (Spigelius, 1578–1625, see p. 163). Galen described about 300 muscles, and several of the names that we now use derive from him, among them the *masseter* and the *cremaster*. For his investigation of muscles Galen used particularly the Barbary ape (*Macacus inuus*, see Fig. 26).

There is reason to suppose that the description of the brain submitted by Galen is less original than his Myology or than his experimental work on the spinal cord (see p. 60). The classification of the cranial nerves in vogue until the seventeenth century was, however, derived from him. In that system the olfactory nerves and trochlear nerves were not recognized. Galen's first nerve was the optic nerve. The oculomotor and abducens were reckoned as the second. The greater part of the trigeminal was the third of Galen's notation, though a part made up his fourth nerve. We may here note that it was not until the time of Meckel (1758) that the anatomy of the trigeminal was adequately known. The facial and auditory nerves were linked together as Galen's fifth cranial pair. His sixth was a combination of our glosso-pharyngeal, vagus, and spinal accessory. His seventh cranial nerve was our hypoglossal (Fig. 28). It is noteworthy that he clearly recognized nervous ganglia and traced the sympathetic system through part of its course. He was, moreover, fully acquainted with the recurrent laryngeal nerves and with the differences in the course that they pursued on the two sides (Fig. 29).

The Angeiology of Galen is less satisfactory than his Osteology and his Myology. He wrote a special work *On the anatomy of the veins and arteries*, but a false doctrine of the movement of the blood prevented the just development of the theme. He was, moreover, ignorant of the process of injection or of any other special method of preparation. The venous system, following hints from the Hippocratic Collection, is compared to a tree of which the roots spring from the abdominal viscera, the trunk is the vena cava, and the branches are to be found in the lungs and in other parts of the body, one of the most important branches being the right ventricle. The veins are also represented as arising in the liver.

A similar description is given of the arteries. The roots of
the arterial system are represented by the *arterial vein*, which
we now call the *pulmonary artery*. The left ventricle and aorta
he regarded as the trunk from which the branches come off.
The arteries, he observed, have walls which are much thicker
than those of the veins. Galen demonstrates that Erasistratus
was in error in thinking that the arteries contain air, and that
blood enters only after incision (p. 32). He does this by a
simple and very effective experiment. An artery is exposed
along a considerable length and ligatured in continuity at
two points. It is then incised between the ligatures. Blood,
not air, flows therefrom. Since blood could not enter through
the ligature, it must have been present before the ligature
was applied.

Galen had a fairly good idea of the general course of the
veins. These structures, he believed, drew nourishment from
the intestines and distributed it to the liver. The vein from
the intestines passed to the liver through the gate (Latin
porta) known as the transverse fissure ; hence our name
portal vein. The close correspondence of veins and arteries
in most parts of the body was well known to him. He knew,
too, the veins of the brain, certain of which still bear his name.

§ 7 *Galen's Physiological System* (*See Fig.* 30)

The basic principle of life in the Galenic philosophy was a
spirit or *pneuma* drawn from the general World-spirit in the
act of respiration. It entered the body through the *trachea
arteria* and so passed to the lung and thence, through the *vein-
like artery*—which we now call the pulmonary vein—to the
left ventricle, where it encountered the blood. But what was
the origin of the blood ?   To this question his answer was
ingenious. It was derived in part from Erasistratus and the
errors that it involved remained till the time of Harvey. Galen
believed that chyle, brought from the alimentary tract by the
portal vessel, arrived at the liver. That organ, he considered,
had the power of elaborating the chyle into venous blood and
of imbuing it with a particular spirit or pneuma innate in all

living substance so long as it remains alive. This pneuma was spoken of as the *natural spirit*. Charged with *natural spirit* derived from the liver and with nutritive material derived from the intestines, the blood, he believed, was distributed by the liver throughout the venous system which

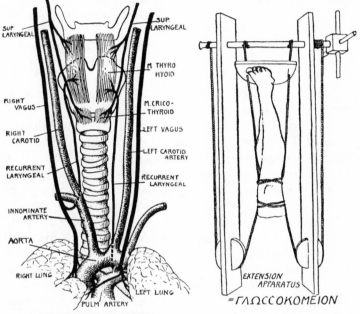

INNERVATION OF LARYNX

FIG. 29.—Course of recurrent laryngeal nerves. Galen compares these in their action to the *glōssokomeion*, an extension apparatus then in use by surgeons. The turning of one screw of this apparatus tightened ropes which pulled in opposite directions. So also he says the two branches of the *Vagus* (the *superior laryngeal* and the recurrent *laryngeal* of modern notation), cause a pull in opposite directions on the laryngeal apparatus. In the absence of an adequate technical nomenclature comparisons of this type were not infrequently resorted to by ancient anatomists.

arises from it, ebbing and flowing in the veins. One great main branch of the venous system was the right side of the heart.

For the blood that entered this important branch, the right side of the heart, the Galenic scheme reserved two possible fates. The greater part remained awhile in the ventricle parting with its impurities, which were carried off by the *artery-like vein*—our pulmonary artery—to the lung, and there exhaled.   These impurities being discharged, the venous blood in the right ventricle ebbed back again into the general venous system. A small portion of it followed a different course. This small portion trickled through minute channels in the inter-ventricular septum and entered the left ventricle drop by drop.   There it encountered the pneuma brought thither from the outside world by the trachea and *vein-like artery* (our pulmonary vein). These drops of blood in contact with the air in the left ventricle became elaborated into a higher type of pneuma, the *vital spirit*, which was distributed through the arteries and with the arterial blood.

Among the arteries some went to the head, and thereby *vital spirit* was brought to the base of the brain. Here the blood was minutely divided by the channels of the *rete mirabile*. In that mysterious organ the blood became charged with yet a third pneuma, the *animal spirit*, which was distributed by the nerves, which were supposed to be hollow.

These three pneumas, the *natural spirit*, the *vital spirit*, and the *animal spirit*, formed the basis of the physiological system till Harvey. The system must be compared to that of Erasistratus (p. 31), who, it will be remembered, recognized only two forms of spirit within the animal body.

Among Galen's most remarkable efforts are the investigations he made of the physiology of the nervous system.   In his treatise *On anatomical operations*, he tells of his experiments on the spinal cord. Injury to the cord between the first and second vertebræ caused, he observed, instantaneous death. Section between the third and fourth produced arrest of respiration. Below the sixth vertebra it gave rise to paralysis of the thoracic muscles, respiration being carried on only by the diaphragm. If the lesion was lower still the paralysis was confined to the lower limbs, bladder and intestines. The physiology of the spinal cord is worked out most ably and in very considerable detail. The knowledge of the functions of the spinal cord was not extended until the nineteenth century, with the appearance of the work of Sir Charles Bell (1811), of Magendie (1822), and of Le Gallois (1830).

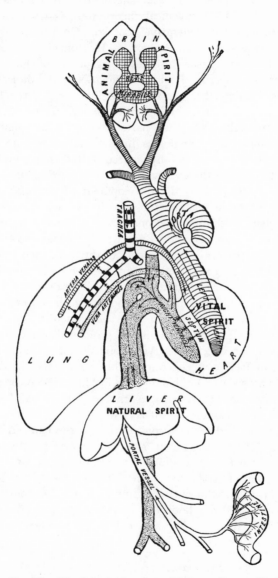

Fig. 30.—Physiological system of Galen. See pp. 58–60.

Very important for the subsequent development of
anatomical and physiological theory were Galen's views on
the nature of the generative process.   On this he differed
very considerably from Aristotle.   He considered, contrary
to Aristotle, that the testicles correspond to the ovaries,
both secreting sperm.   Basing his remarks on the dissection of
animals, he describes the uterus as bifid, each branch conveying seed from the corresponding ovary.   These cornua of
the uterus, like the other parts of the female generative
system, have their analogies in the male anatomy.   The
human uterus was generally figured as bicornuate until the
end of the sixteenth century.

Galen's ideas as to the development of the embryo are
peculiar, being less satisfactory and less based on experience
than is usual with him.   He considers that the first organ to be
formed in the embryo is the liver, which is congealed from the
blood, next the brain is formed from the seed, then the heart
is formed from the blood.   A little later blood-vessels and
nerves are also formed from the seed.   These ideas also recur
in literature until the middle of the sixteenth century.

Galen's writings brought him prominently before the
scientific public, and gave rise to prolonged controversy, in
the course of which he gave public demonstrations in the
Temple of Peace.   A collection of his works was stored there
and, towards the end of his life, was lost in a fire. As he had no
copies of some of these writings, the loss was irreparable.   The
bulk of the remainder is still very impressive.

Galen retained his position of trust to the end.   After the
death of Marcus Aurelius in A.D. 180, he became the adviser
of Commodus (A.D. 161–192), on whose death, in A.D. 192, he
was appointed physician to the Emperor Septimus Severus
(A.D. 146–211), who outlived him.   His writing remained in
standard use throughout the Middle Ages and on into the
sixteenth century.   He was habitually spoken of as the
" Prince of Physicians ".

§ 8   *The Dark Ages, about* A.D. 200–1050

Galen established no school, nor had he any definite
followers.   His character and love of controversy was not of

the kind that would endear him to disciples. On his death in 199 the active prosecution of anatomical and physiological inquiry ceased absolutely. The curtain descends at once, and for the subject we are discussing the Dark Ages have begun.

Anatomy in the pagan world descends into darkness more abruptly, but not more surely, than Philosophy. The whole system is soon to be overwhelmed. Alexandria has long been in decline ; a mob, fanatically Christian, has destroyed her school and library, with all the hoarded wisdom of the pagan past. Men of the new faith fix their eyes on the wrath to come and the glory after it. In the race for salvation, who will pause to consider this miserable tenement of clay ? The barbarian is at the gate. The Empire falls in smoking ruin. After the fire a flood ; wave on wave it breaks, Vandal, Goth, Lombard, Hun, Saracen, and Northman. The hand of the Lord is heavy ; His day is surely nigh. A pause, and at last the waters subside. The Church, the captive daughter of Zion, arising from the ruins, gathers around her the scattered remnants of mankind. She knows well the strait path for those that would be saved. In that assurance men may turn to examine the wreck of their world. What is there left of ancient Science ? A few peaks have been but awash with the flood, and not wholly overwhelmed by it. In the South of Italy, especially, Salerno has been least injured. There Greek still lingers, and even some remnants of the ancient writings. Messengers come from the East, where the flood of Islam has rolled back from the barbarian watershed. They tell that there, too, something has been saved. The relics of ancient wisdom must be salved, at least for the healing of our bodies. The Scholastic Age has begun.

These ages of disorder was associated with a steady process of decay in the amount and accuracy of available medical knowledge. So far as the West was concerned, the human intellect touched nadir about the end of the tenth or beginning of the eleventh century. In this deterioration Anatomy was involved with the other Sciences and Arts. With the external and political conditions associated with this progressive mental paralysis we are not here closely concerned, but we may allude to two purely intellectual factors in the process.

First was the wide acceptance of the Christian doctrine that the body was of little or no import in comparison with the soul. Men were so certain of the existence, the word, and the will of God that Galen's teleology was for a time almost forgotten as irrelevant and unimportant. What need to prove the things one knew ?   The body being contemptible, was unworthy of study and Anatomy was the most vain of all those empty pagan sciences that did but concern themselves with the external temporary and perishable world. In that day of wrath, that dreadful day, when Heaven and Earth shall pass away, when shrivelling like a parched scroll, the flaming heavens together roll, what then can or will avail these pitiful details of Anatomy ?   The mind of the mediæval man was very closely set upon his end.   Death and the things that come after was an obsession of the age. No pains were spared to keep death always in memory.   The memento of death is very characteristic of the Middle Ages. Figures of skeletons and decaying corpses are not only to be found on tombs, but on finger rings and house ornaments, in manuscript illuminations, and elsewhere.

But, besides this negative and deterrent influence of Christian teaching was another and positive doctrine that not only turned men from the study of their bodies, but inculcated a fundamentally false conception of the nature of those bodies.   This fallacious view, to which we have already alluded, was no integral part of Christian teaching, but can be traced back into Greek philosophy for centuries before Christianity had appeared on the scene.   It is the old idea that the human frame foreshadows the structure of the Greater World, that the *Microcosm* is built on the same model as the *Macrocosm*, that Man is an epitome of the Universe (see p. 15).   It is a view that easily allied itself with, even if it did not spring from, the body of astrological doctrine that first Greece and then Rome derived from Babylon.

The whole astrological system had become elaborated in the Early Christian centuries ; with philosophical thoroughness by the Stoics ; with mystical intent by the Neoplatonists, by the Gnostics, by the Mithra worshippers, and by the other sects that were still the competitors

PLATE VIII

Painted wooden Graeco Egyptian Sarco-
phagus of about the time of Christ, now
in the British Museum. Around the
figure the signs of the zodiac are written.
These are supposed to influence the life
of man, a tradition passed on to the Middle
Ages. Compare Plate XVI.

[face p. 64

of the religion that was to displace them all. By these sects and by those which sprang from them not only were the regions and luminaries of the firmament held to influence the parts of man's body and the course of his life, but it was also maintained that the supposed connexion between constellations and corporeal organs had a spiritual significance and a moral interrelationship. These ideas, at first resisted by the fathers of the Church, were at last absorbed by those who professed the religion that ultimately prevailed. Thus was disseminated a whole mass of vapid and speculative belief which, if not a part of Christian teaching, became at least part of the teaching of Christians, and accepted by the leaders of the Church.

Among the most common products of these ages is what seems now a childish scheme in which the signs of the zodiac are written first around and ultimately upon the various parts of the body that they were thought to govern. Such schemes may be widely traced in the tombs of Græco-Roman Egypt (Plate VIII), among the military monuments of the later Empire, in the pious documents of the monkish period that preceded the great intellectual revival of the twelfth and thirteenth centuries or in the handbooks of the barber surgeons of the Renaissance period (Plate XVI). The *melothesia* or *zodiacal man*, as the scheme was called, permeated these centuries. It is among the commonest of mediæval diagrams. Belief in it superseded Anatomy and Physiology. The universal faith in astrology is itself enough to explain the decay of these studies.

III

THE MIDDLE AGES AND RENAISSANCE,

1050–1543

§ 1 *The Translators from the Arabic, about* 1050–1250

A FTER Galen we encounter no anatomical activity for many centuries. A few later Greek writers exhibit the Galenic tradition in more or less corrupted form. A few others

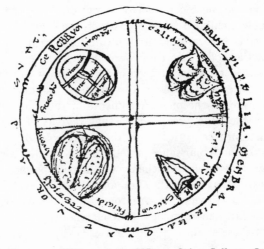

FIG. 31.—From an XIth century MS. at Caius College, Cambridge (MS. 428, folio 50). Around the diagram is written *Principalia menbra virilia quatuor adsunt,* " There are four principal human members." Within the circle are figured *Cerebrum* (N.W.) ; *Epar* (N.E.) ; *Cor* (S.E.) ; and *Testiculus* (S.W.). Each of the organs is associated with two of the primary " qualities "—thus the cerebrum is cold and moist (*frigidus, humidus*) and the heart is hot and dry (*calidum, siccum*) and so on (see Fig. 16). The brain is shown divided into three sections marked *fantasia, intellectus,* and *memoria* (see Fig. 41).

show some originality in presenting their views, but no further direct access to the facts. Among the Latin-speaking peoples of the West the anatomical tradition of antiquity was even more corrupted. Beyond astrological diagrams, such as those to which we have already referred (Plates VIII and XVI), we get only the most childish misinterpretation of ancient doctrines (Figs. 31, 32). We shall therefore not take up our tale until the human mind begins to rise from its secular depression.

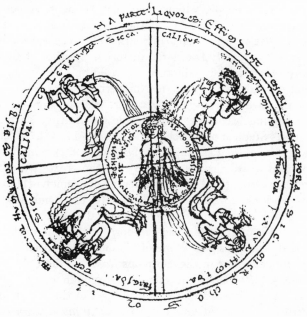

FIG. 32.—From the same XIth century MS. (folio 23 verso) as the previous figure. Four angels are shown pouring from vessels the four humours into the body of man. Around is written *Quatuor humores bisbina partes* (?) *liquores Effundant teneri per corpora sic microchosmi*. The four surrounding figures typify the four humours which are associated with the four primary qualities in the same way as are the four principal organs in Fig. 31.

Intellectual leadership passed about the eighth century to people of Arabic speech and remained with them till the thirteenth century. Thus it came about that the most important documents of Greek medicine were translated into Arabic. Translations of these originally Greek works from their Arabic dress into Latin formed the main mass of scientific reading in the West for long after the process of reawakening.

The first recovery of medical material from these Arabic sources took place in the eleventh century at the monastery of Monte Cassino in South Italy. Here the monk Constantine, the African (died 1087), himself an ignorant and dishonest worker, began his series of translations from the Arabic. Dating from about 1100, we have two descriptions of the dissection of the pig which have been thought by some to indicate anatomical activity at the earliest of all the medical schools, that at Salerno, also in South Italy. These documents, however, are merely translated material. Neither at this nor at any other time was dissection carried on at Salerno. The actual revival was, in fact, in Northern Italy.

In the twelfth century several translators were at work rendering medical works from Arabic into Latin ; earliest of them was Stephen of Antioch, who produced in 1127 a version of a treatise of the Persian, Hali Abbas (died 994), containing an important anatomical section. By far the greatest of all the translators from the Arabic was Gerard of Cremona (1115–85), who, working at Toledo, rendered into Latin no less than ninety-two works, many of philosophical value. The figure of Gerard is of great historical import, in that his work made Scholasticism possible. None of his medical translations was more influential than the enormous *Canon* of the Bokhariote, Avicenna (980–1037), the anatomical section of which was the most widely read text on the subject in the Middle Ages. Gerard also rendered into Latin the anatomical work of the Persian, Rhazes (died 932). It is to these three Arabic writers, Avicenna, Hali Abbas, and Rhazes that the main mass of medical knowledge before 1500 can be traced. All three writers, so far as Anatomy is concerned, themselves depend on Arabic versions of Galen. Several of the works of Galen and Hippocrates were also translated from Arabic versions in the thirteenth and fourteenth centuries.

The twelfth and thirteenth centuries, though barren in anatomical achievement, exhibited enormous mental activity. The scholastics, though they very seldom stooped to observation, were frequently of a speculative turn of mind. Dating from the twelfth century, we have a series of mystical anatomies precipitated with more or less clearness from the old theory of Macrocosm and Microcosm (p. 65). These are based, mediately or immediately, on Arabic material. With the thirteenth century we are moving in a more positive atmosphere. In the person of Albertus Magnus (1206–80) we encounter a genuine naturalist. Albert's knowledge was drawn from Arabic-Latin versions of Aristotle. He did at times, however, observe for himself and has left us some account, for instance, of the development of birds and of fishes. The movement which he represents was to come to fruition, so far as our subject is concerned, in the following century.

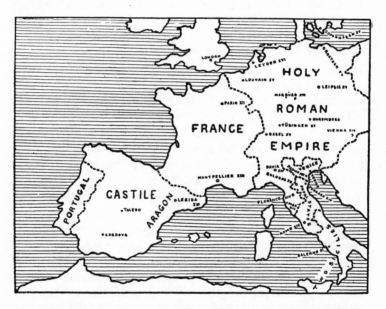

Fig. 33.—Map showing the centres of anatomical study during the Middle Ages and Renaissance. In the case of towns possessing Universities, the century of foundation of the University is given in Roman figures.

§ 2 *The Rise of the Universities (Fig. 33). The Bologna School*

In the great awakening of the thirteenth century, a large part was played by the Universities. These were established in numbers during the century, and university life gradually came to exercise a profound effect on social, political, and intellectual conditions.   In most of the Universities Medical Faculties grew up and in many Anatomy came to be studied. The texts used were those of Avicenna, of Hali, and of Rhazes. But the atmosphere of the Universities was utterly scholastic and Scholasticism, however it may sharpen wits, does nothing to develop the senses.   Observation of Nature was wholly neglected.   There was no practical anatomical instruction. The statements concerning public dissections at the Universities during the twelfth and thirteenth centuries are due to misunderstanding. The actual anatomical achievements of the thirteenth century were exclusively of a literary character.   There was a multiplication and improvement of the Arabic-Latin texts and the recovery of a few from the Greek.   The first credible witness of actual *public* or semi-public dissection that reaches us is from about the year 1300, and the place where we get news of it is Bologna.

The school of Bologna is of very great antiquity, and is perhaps the most ancient of the institutions to which the term University can be rightly attached.   An organized Medical Faculty existed there as early as 1156.   The teaching at Bologna, as in other medical schools, consisted entirely of readings of Latin translations from the Arabic which were becoming ever more accessible.   As the Universities multiplied they began to some extent to " specialize ". Bologna appears first in history as a Law School, and continued to develop along the same line.   In the second half of the thirteenth century Bologna was by far the most important seat of legal learning in Europe.   For long the Medical Faculty there was directly dependent on the Jurists and not till 1306 was it even permitted to elect its own head.   There can therefore be no doubt that the opening of the human body at that University began in the first instance with full knowledge of the lawyers who were in supreme control.   It therefore seems highly probable that it began as a part of a forensic process.   The first post-mortems held at Bologna were probably conducted towards the end of the thirteenth century.

§ 3 *The Beginning of Dissection,* 1250–1300

The early advent of dissection has often impressed the historian. It is not easy to find evidence of Nature-study in the thirteenth or fourteenth century, nor of any enthusiasm for putting theoretical considerations to practical tests in that age. There was still no Botany worthy of the name, no Zoology, hardly any naturalistic Art, no experimental Science, no systematic record of observation in any department. Yet dissection had become comparatively common at Bologna by the end of the first quarter of the fourteenth century. The question is asked why men, obviously so little interested in Nature and Nature's ways, should have bent themselves to so repellant a process as dissection of the human body in order to seek out the secrets of Nature ? The answer is that, in fact, they did nothing of the sort. Dissection in the fourteenth century did no more, and was asked to do no more, than verify Avicenna, —whom nobody doubted. It seems probable that the earliest reason for examining the human body was simply the gathering of evidence for legal processes. This is a reason, and the only reason, that would have appealed to an official of Bologna University of the thirteenth century. As time went on post-mortem examination passed into anatomical study.

We may investigate the actual history of the subject so far as it can be put together from the sparse records. The founders of the Surgical School at Bologna were Hugh of Lucca (about 1170–1240) and his pupil, the cleric Theodoric Borgognoni (1205–98). The work of the latter has survived, and contains, imbedded in it, descriptions of methods adopted by Hugh. Theodoric's surgery borrows its Anatomy direct from the Arabians, and contains no evidence for dissection. The opportunities of Theodoric for surgical practice at Bologna ceased in 1266, when he became Bishop of Cervia, and went to live at Lucca. This is the *terminus a quo* for the beginning of dissection at Bologna. Had it been practised there before 1266 it would have reacted on Theodoric's work.

William of Saliceto was a Bolognese surgeon (1215 ?–1280 ?)

who became a teacher in his own University. He left a very able treatise on Surgery, containing a section on Anatomy. The anatomical portion is borrowed from the current Arabian anatomies, and does not mention the practice of dissection. Nevertheless, it contains some evidence of direct access to the dead human body. Thus the arrangement is different from that of its Arabian sources, and is more natural and in accord with the actual disposition of the parts. Moreover, the work reads like that of one who had at least seen dissection. Thus he speaks, for instance, of the appearance of the intra-thoracic organs of a wounded man. The view could only have been obtained after death. William of Saliceto's Surgery was completed in 1275, and that year is a *terminus ad quem* for the practice of dissection at Bologna. We may thus place the beginning of anatomical study in the decade between 1266 and 1275.

A most interesting contemporary of William of Saliceto was Thaddeus of Florence (1223–1303), who also taught at Bologna. This very remarkable man perceived the importance of access to Greek sources, as distinct from Græco-Arabic, and he encouraged the preparation of good Latin translations of medical works direct from the Greek. He stamped his personality on the whole development of Medicine at Bologna, and he is bound up with the beginning of dissection in a peculiar way. Not only does he give occasional hints of post-mortem examination in his works, but all the first genera-tion of writers who refer to the practice—Bartolomeo da Varignana (died 1318, see p. 73), Henri de Mondeville (died 1320, see p. 73), and Mondino de' Luzzi (died 1326, see p. 74 ff.)—were his pupils. It is quite certain that before the death of Thaddeus post-mortem examination was being openly performed at Bologna. As one investigates the revival of Medicine in North Italy the lines always seem to converge on him. There is a mass of unpublished material concerning this extraordinary man in the Vatican Library, and its investigation may throw light on the origin of dissection.

The first frank reference to post-mortem examination is from the year 1286. The Franciscan Salimbene of Parma (1221–90 ?), in his chronicle written in 1288, tells that a pestilence raged in Italy in 1286, and that a physician of Cremona then opened a corpse to see if he could find the cause of the disease. It appears that he opened only the thorax to glance at the heart. The first formal account of definite post-mortem examination is fourteen years later. In February, 1302, a certain Azzolino died at Bologna under suspicious circumstances. Poison was suspected, a judicial inquiry was held, and a post-mortem examination actually ordered by the court. The investigation was conducted by two physicians and three surgeons, with Bartolomeo da Varignana at their head. Their report is in existence, and terminates with the words : " we have assured ourselves of the condition by the evidence of our own senses and by the anatomization of the parts." The statement is given without any remark that the proceeding was unusual. This Bartolomeo da Varignana was a pupil of Thaddeus and was professor of Medicine at Bologna, an office in which he was succeeded by his son and by two of his grandsons. An actual contemporary illustration of a post-mortem has survived, and is in the Bodleian Library (Fig. 34). The picture may be dated at about 1300, with a not impossible error of twenty to thirty years in either direction. It shows a surgeon opening the body and extracting the organs in the presence of a physician and of a monk. It is of Norman or Anglo-Norman workmanship.

We have another interesting document bearing on the early practice of dissection at Bologna. At the very end of the thirteenth century, and after the death of William of Saliceto, there came there a Norman student named Henri de Mondeville (about 1270–1320, Plate IX). He profited by the instruction of Thaddeus and the other professors and then left. He sojourned for a while in Northern France and Flanders. Before 1301 he settled at the famous Medical School at Montpellier. Here, in 1304, he was giving anatomical lectures. These lectures were illustrated by large diagrams, to which he refers in the section on Anatomy prefixed to a work on Surgery that he wrote for his students. It is probable that he brought these diagrams with him from his old school at Bologna. The originals have disappeared, but a copy made in 1314 is now in the *Bibliothèque nationale* at Paris. In the copy the diagrams are greatly reduced in size, and the artist, who was a Norman or a Fleming, was obviously ignorant of Anatomy. Nevertheless, his little pictures give an indication of the manner of dissection at Bologna at that time (Plate X). The anatomical text of Henri de Mondeville is mainly borrowed from Avicenna and the Arabs.

Fig. 34.—Miniature from a MS. in the Bodleian Library at Oxford of the early XIVth century (Ashmole 399, folio 34). It is the earliest known representation of dissection. The operator, who is in layman's dress, is being addressed by a physician and a monk. The body, that of a female, has been opened from xiphisternum to symphysis pubis. Kidneys, heart and lungs, stomach, etc., are strewn around. The operator holds the liver in his hand.

§ 4   *Mondino, the Restorer of Anatomy*, 1300–25

For the next worker we are in the full light of history. Anatomy has become a recognized discipline, and we are no longer in doubt as to how it is practised. Mondino de' Luzzi (*c*. 1270–1326) was born at Bologna, and studied in his native city. He, too, was a pupil of Thaddeus, and he was a

PLATE IX

HENRI DE MONDEVILLE

engaged in lecturing. From a French MS. of his *Surgery*
written in 1314 and now in the Bibliotheque nationale at
Paris (Fr. 2030).

fellow-student of Henri de Mondeville. He graduated about 1290, and joined the teaching staff of the University in 1306. He systematically worked at Anatomy and dissected the human body in public.

The *Anothomia* of Mondino, written in 1316, is the first modern work on the subject. Those who preceded him incorporated their anatomical work in larger treatises on Surgery, and do not refer directly to their own anatomical experiences. With Mondino this is changed. His work is entirely devoted to Anatomy, and is essentially a practical manual of the subject. Mondino is with justice called the " Restorer of Anatomy ". Let us glance at his book.

We note first his involved constructions and his debased Latin. Next there comes before us his very confusing nomenclature which makes his book extremely difficult to read. On the one hand, for many parts he has several names. Thus the sacrum is variously described as *alchatim*, *allannis*, and *alhavius*. On the other hand, the same name is often used for several parts. Thus the word *anchæ* may mean the hips, or the pelvic skeleton, or the acetabulum, or, again, the corpora quadrigemina. Thirdly, we remind ourselves that, in the absence of preservatives, dissection had to be performed hurriedly, and this especially with the abdominal viscera Work was sometimes continued through the night, and the whole process completed in four days, a day each being given to the belly, the thorax, the head, and the extremities in that order. This haste has left its mark upon the book. Fourthly, we recall that subjects for dissection were not easy to come by. They were normally criminals. Even as late as the sixteenth century criminals were sometimes executed in a manner chosen by the anatomists. There is ample internal evidence in the book of limitation of material. Fifthly, we observe that, though dissecting on his own account, Mondino is yet relying almost entirely on Arabian authorities. He is really dissecting to memorize their works, much as a student nowadays dissects to memorize his textbook, not to enlarge knowledge nor to make discoveries. The scientific spirit has hardly awakened in Mondino. Nevertheless, he has made a great step forward.

But lastly, and above all, we would emphasize the fact that Mondino dissected *in person*. In this respect he was wiser or more courageous than most of his successors until the time of Vesalius.   As dissection gained formal inclusion in the curriculum, the professor became further removed from the object of his study.   He literally " rose with his subject ". Leaving his position by the cadaver (Fig. 34), where he might demonstrate to his students, he ascended his high professorial chair.   The " chair " of a professor was very much of a physical reality in those days ; a great elevated structure provided with steps and a reading desk, something like a pulpit (Fig. 35).   From there he read or lectured while a junior colleague or *ostensor* pointed out the line of incision and a menial *demonstrator* performed the actual dissection.   All was thus done third hand and according to the written word.   We are in the scholastic period and must not expect any frequent appeal to Nature.   Having once got into his chair, the professor did not willingly descend from that dignified position.   Nor was it from the University that the second great reform in Anatomy was to come in the centuries which followed.   The second reform came, as we shall see, from quite another quarter, the Studio.   Thus it is saying a very great deal for Mondino that he was his own demonstrator.   He took the first and perhaps the greatest step.   It was two centuries and more before the next was taken.

Mondino had read widely among the Arabian anatomists, and naturally borrows from them.   Nevertheless, his work contains a considerable number of references to actual anatomical procedure.   Moreover, he deals not only with Anatomy in our modern sense, but also includes physiology and much discussion of the application of anatomical and physiological principles to Medicine and Surgery.   His book thus gives a good deal of insight into the scientific knowledge of the day.

The *Anothomia* of Mondino is not arranged like a modern anatomical textbook, which deals with the various systems consecutively.   It more nearly resembles a modern manual of dissection, in which the organs are described in the order in which they present themselves.   Thus, after discussing the

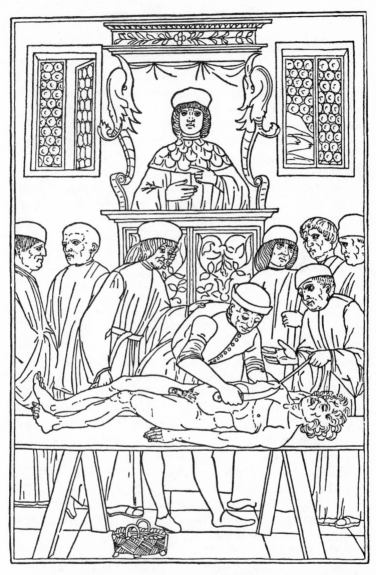

Fig. 35.—Dissection scene from the Italian *Fasciculo di Medicina* printed at Venice in 1493. The picture forms the first page of an Italian translation of the *Anothomia* of Mondino. The professor in his robes is reading his lecture from his chair. A menial *demonstrator* dissects, as directed by the wand of the *ostensor*. Students in academic dress stand around and look on, but do not themselves dissect.

scope and manner of the inquiry, he opens with the *natural members*, that is the parts associated with digestion. The abdominal wall is described in some detail, and, that completed, Mondino passes to the alimentary canal and then to the spleen, liver, and great vessels of the abdomen. The generative organs come next, and the description of them is very full. He now turns to the *spiritual members*—the thorax and its contents—described, like those of the abdomen, from without inwards. Now follow the *animal members*, that is, the parts of the head, and these are treated very systematically, proceeding from the scalp and going deeper till the base of the skull is reached. Lastly, the work describes the spinal column and the structures surrounding it, together with the extremities. These are treated in a very superficial and perfunctory fashion.

§ 5   *Mediæval Anatomical Nomenclature*

It is appropriate that we should here turn aside to consider the anatomical language of Mondino. The names of the *natural*, that is the abdominal, members will at once strike the reader. These are, of course, in relation to the organ of the *natural spirit*, the liver to wit. At the very beginning we encounter such strange forms as *Mirach*, i.e. anterior abdominal wall ; *Siphac*, i.e. anterior layer of peritoneum ; and *Zirbus*, i.e. great omentum. These are part of the anatomical nomenclature of the day, derived, not, as is ours, from Greek and Latin, but from Arabic. For long this Arabic system remained in use. With the revival of Greek in the fifteenth and sixteenth centuries, these terms gradually gave place to others of classical origin.

Some of Mondino's obsolete terms that have followed the Arabic into oblivion were not, however, derived from Arabic, but were of classical origin, and had been in continuous employ right through the Dark and Early Middle Ages of Anatomy. Such was, for instance, the word *longaon*, i.e. rectum, a word which survived in English usage until the eighteenth century. Of these old words of classical origin,

Mondino occasionally gives quite false etymologies, as, for instance, when he says that *Colon* is derived from the (mediæval) Latin word *cola*, i.e. cellule. Colon is, in fact, a Greek word which meant primarily an *organ*, though it had acquired its present meaning at least as early as Aristotle and occurs in the same sense in Latin in Pliny (A.D. 23–79), whose work on *Natural History* was very widely read during the Middle Ages. These words of classical origin are, however, in a minority, and most of Mondino's nomenclature is derived from Arabic.

During the sixteenth century a regular warfare was waged on behalf of the Latin and Greek as against the Arabic anatomical terms. It is difficult for us now to understand the virulence imported into the discussion. These unfortunate words were regarded as symbols of two cultures, the Arabist and the Humanist. To add to the confusion, just as the Humanists were gaining the upper hand the new experimental Anatomy came in. The exponents of the new science became nearly as violent against the Humanists as the Humanists themselves had been against the Arabists. Looking back from the vantage ground of Time we can see that without the Arabists the human mind could not have been raised out of the slough of the Dark Ages. Without the Humanists the anatomists of the sixteenth century would have had to begin on a much lower rung of the ladder of knowledge than that at which in fact they started.

Considering how persistently anatomical nomenclature has been " purified " from Arabic terms, it is remarkable that any should have survived. Yet our textbooks are still employing a number of them. This in itself shows that they were not inapt and that they fulfilled a real purpose. It must be remembered that the surviving terms have been carefully Latinized and Græcized since the time of Mondino, and neither an Oriental nor a Classical scholar will easily recognize them for what they are. It is a case of protective mimicry in the world of words ! Of all our surviving Arabic anatomical terms, the oldest is *Nucha*, a word introduced by Constantine the African about 1080. Others of Arabic origin were derived from Gerard of Cremona's Latin translation of Avicenna made about 1180. From this version of Avicenna come, for instance, Basilica (for the vein has nothing to do with the Basilica of architecture), Cephalica (which is not derived from the Greek root *cephalic*, though that, too, is largely employed in Anatomy), Retina (which is unrelated to *rete*), Saphena (which is not from the Greek *saphēnēs* = clear, evident), and

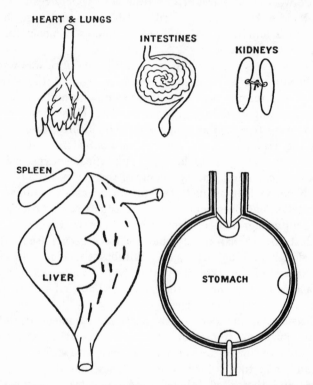

Fig. 36.—Diagrams of the bodily organs from a MS. in the Bodleian Library at Oxford of the early XIVth century (Ashmole 399, folios 23 and 24). Note the spherical stomach, the five lobed liver and the slipper shaped spleen.

Sesamoid (which is the " open sesame " of the story of Aladdin).

In addition to such Arabic words which have clung on in our anatomical vocabulary, there are others expressing Arabic ideas. The mediæval Latin translators from Arabic were very

PLATE X

A B

C D

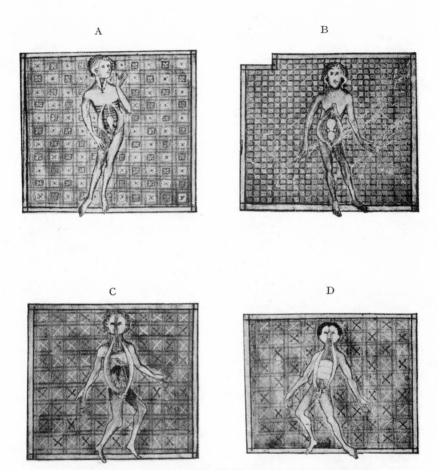

ANATOMICAL DIAGRAMS OF HENRI DE MONDEVILLE

From a French MS. of his *Surgery* written in 1314 and now in the Bibliothèque nationale at Paris. A and B represent the male and female urinogenital systems respectively. C and D represent general dissections exhibiting the viscera. The peculiar pose of D should be compared to that of the Vesalian tabula exhibited in figure 109 where it is shown reversed.

literal and their renderings are thus recognizable in our modern terminology. Thus to Gerard's translation of Avicenna we can trace *Clavicle* and *True* and *false* ribs, to Gerard's translation of Rhazes we owe *Albugineus* and *Iris*. Stephen of Antioch's translation of Hali Abbas has given us *Pia mater* and *Dura mater*. Most of these surviving terms of Arabic origin were originally popularized by Mondino, who seems, on his own account, to have put into circulation the words *Matrix* and *Mesentery* in their modern anatomical connotation.

§ 6 *Anatomical Knowledge of Mondino*

Mondino's description of the organs is naturally very elementary and often inaccurate. Thus, the stomach, as is usual in mediæval works on Anatomy, is described as spherical (Fig. 36). The liver has five lobes, a very persistent idea taken from the anatomy of the dog (p. 43). Great and special attention is paid by Mondino to the gall bladder, the seat of one of the humours, the *Yellow Bile* (Choler). The *Black Bile* (Melancholia) is secreted by the slipper-shaped spleen (Fig. 36), and evacuated through imaginary channels into the cardiac end of the stomach (Fig. 37). The cœcum is described without any vermiform appendix. The description of the pancreas is very obscure, though, oddly enough, its duct is referred to ; this duct, it is usually considered, was first described by Wirsung (died 1643), a pupil of Vesling (1598–1649) at Padua about 1641. The urine is represented as literally " filtered off " by the kidneys from the blood, an idea taken from Galen and still adhered to by Vesalius.

The detailed description of the generative organs by Mondino is worth some attention. He notes the different origin of the spermatic vessels on the two sides. The uterus is divided into seven cells (Figs. 38, 39), a conception that he must have culled from the writings of that muddle-headed magician Michael the Scot (about 1178–1234). In an interesting passage he seeks the analogies between the male and female generative organs, and in this matter his conclusions are not widely different from those in the first

anatomical works of Vesalius more than two hundred years
later (Figs. 61, 62, p. 112). As regards the physiology of
generation, he halts between the views of Aristotle and of
Galen. He has a very good passage describing the operation
of hernia both with and without castration. Oddest of all
is his description of the treatment of an incised wound of the

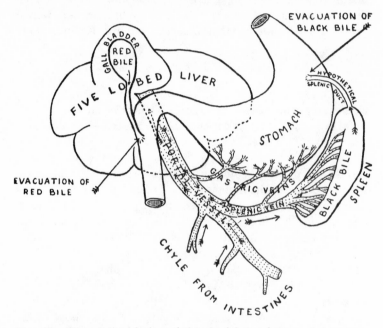

FIG. 37.—The mediæval view of the physiology of the portal system.
The portal vessel brings chyle from the intestines to the five
lobed liver. Red bile is evacuated from the liver by the bile
duct into the duodenum. Blood is sent from the liver to the
spleen via the splenic vein. Black bile (melancholy) is evacuated
into the stomach either by a hypothetical duct near the cardiac
end of the stomach or by the gastric veins.

intestines. The edges of the wound should be closed by
making ants bite on them and then cutting off their heads !
He has a good description of the operation of cutting for
stone.

In his account of the chest, he shows that he had access
to a translation of Aristotle's *Parts of animals* made direct from

the Greek. Such a version had been prepared about 1260.
As his master, Thaddeus, was particularly interested in
translation direct from the Greek, we may suppose that it
was from him that Mondino received the tradition. We note
that another pupil of Thaddeus, Bartolomeo da Varignana,
left some translations from the Greek to the University of
Bologna when he died in 1318. In his description of the heart
borrowed direct from Avicenna, Mondino lapses into the
worst Arabian standard. He describes three ventricles, the
third, a middle ventricle, lying in the thickness of the septum.
This is an attempt to reconcile the accounts of Aristotle and
of Galen. This mysterious organ was still represented in
editions of Mondino that were being printed in the sixteenth

Fig. 38.—Tracing from an illustration in a fourteenth century MS.
of Michael Scott, *De secretis naturæ*, exhibiting seven-chambered
uterus. See C. Ferckel, *Archiv für Geschichte der Medizin*, x,
p. 262, Leipzig, 1916.

century (Fig. 40). Mondino's description of the larynx and
epiglottis also is very confused. The œsophagus is described
under its Arabic term *meri*.

There are many noteworthy points in Mondino's description
of the head. As regards the cranial nerves, he relies on Galen's
Uses of the parts of the body of man, of which a Latin
abbreviation of the Arabic was available. The cavity of the
brain is divided into three vesicles or ventricles. The anterior
is double and is the meeting place of the senses. It is thus the
sensus communis (Fig. 41). The mediæval anatomical
use of these words has given rise to our modern phrase
common sense. The middle vesicle is the seat of imagination.
The hindermost is associated with memory. The scheme

is a commonplace both among Easterns and Westerns throughout the Middle Ages. Mental operations are controlled by the movements of a *red worm*, our choroid plexus (Fig. 41), which opens or closes the passages between the ventricles. These crude ideas as to the physiology of the brain survived to the sixteenth century and are frequently represented in early printed books (Fig. 41). Although Mondino places thought and sense in the brain, he maintains also the old Aristotelian view that the brain cools the heart. Mondino's general physiology is, however, that of Galen, from whom he derives his idea of the three orders

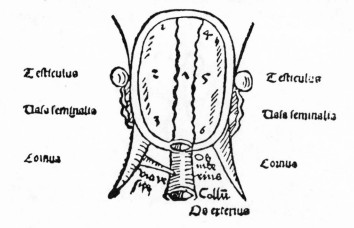

FIG. 39.—Figure of the uterus from the *Antropologium* of Magnus Hundt, Leipzig, 1501. The legends read, on either side *Testiculus, Vasa seminalia, Cornua,* and below *Via vesice, Os matricis, Collum, Os exterius.*

of spirit, *natural, vital, animal* (Fig. 30). His description of the globe of the eye places the lens in the centre, as did Galen, followed by all the Arabs, and all their successors (Plate XII) till Columbus (p. 140) and Plater (p. 133). He describes the operation of couching for cataract.

Mondino's description of the extremities is utterly inadequate. He tells us, however, that he is preparing another work on them, though nothing of the kind has survived. He refers to preparations formed by cleaving the body in the sagittal plane and representations of such preparations have come down to us.

A word should be said about Mondino's anatomical methods. In addition to ordinary fresh dissection, he used preparations dried in the sun. These were supposed to exhibit the general direction of the tendons and ligaments. Mondino tells us that it is very difficult to trace the nerves to their destination, and to this end he examined macerated bodies. The method of maceration was still in vogue in the time of Vesalius, who illustrates it in one of his humorous historiated initials.

Mondino exhibits reluctance to clean bones completely and says he will not boil certain bones *owing to the sin involved*

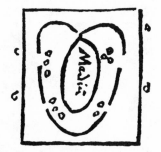

Fig. 40.—Diagram of the heart from an edition of the *Anothomia* of Mondino published at Strasburg in 1513. The diagram shows the three ventricles of which the middle is marked *Medium*. Valves are obscurely indicated at the four orifices in the two " lateral " ventricles.

therein. This passage has given rise to considerable controversy, but the meaning is clear. There was in the Middle Ages a custom of boiling the bones of distinguished persons who died far from home. This was done that their bones might be laid to rest in the place that they had chosen. To prevent this, Pope Boniface VIII issued in 1300 a famous bull excommunicating those who followed the practice. The bull was not directed against the anatomists, but it told against them. This we know not only from Mondino, but from other anatomists of his century. Thus, in 1345, Guido de Vigevano produced in France an anatomical text illustrated by figures which show the actual process of dissection (Fig. 42). He opens

86 DISSECTION AND THE CHURCH

by explaining that the Church prohibits dissection. It is right
to add that a brief of Sixtus IV (Pope 1471–84), who had been
himself a student both at Bologna and at Padua, recognized
the opening of bodies conditional on permission of the
ecclesiastical authorities. Doubtless this was a factor in the

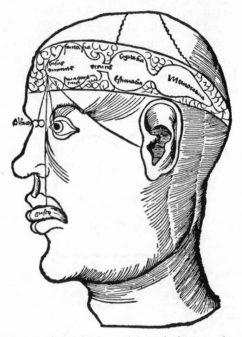

FIG. 41.—Diagram of the brain and its relations to the senses and
intellectual processes from G. Reisch, *Margarita philosophiae*,
Freiburg, 1503.   On the brain are written the following legends :—
fantasia   *cogitativa*
sensus
communis   *vermis*   *memorativa*
         imaginativa   *estimativa*
These occupy positions on the three traditional ventricles except
the *vermis*, which is between the first and second ventricles.   To
the *sensus communis* converge nerves from the ear, eye, nose
(*olfactus*) and mouth (*gustus*).

development of Anatomy at the end of the fifteenth century.
There is ample evidence of occasional dissection before that
time, but from then onward the supply of corpses for dissection,
though limited, was fairly regular. The practice was confirmed
by Clement VII (Pope 1523–4).

§ 7 *The Later Middle Ages, about* 1325–1500

From the thirteenth until the sixteenth century the history of Anatomy remains largely in the custody of the Bologna School. At Bologna dissection received official recognition in

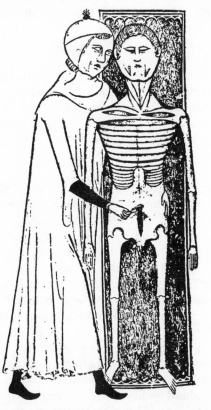

FIG. 42.—Dissector at work from a MS. of Guy de Vigevano at Chantilly written in 1345.   The relation of the dissector to the subject in this and other miniatures in the same MS. shows that the body was suspended.

the University Statutes of 1405 and the same event took place at Padua in 1429.   Throughout the fourteenth and fifteenth century, however, dissection was going on at both places. Dissection was also early practised at Venice, though it was not a University town, and at a few other Italian centres.

Beyond the Alps the most important medical schools were at Montpellier, where public dissections were decreed in 1377, and at Paris, where they were instituted in 1478.

At Bologna Mondino was succeeded in his Chair by a series of workers who made no contribution to anatomical knowledge. They followed the scholastic way in lecturing from a book in their professorial chair, without approaching the body. With their names we shall not burden the reader. A pupil of the Bologna school, the French surgeon, Guy de Chauliac (1300–70), was very influential in standardizing surgical practice, especially in France and England. Nevertheless, his Anatomy is the weakest part of his work, and exhibits little of the practical dissector, though there can be no doubt that Guy had assisted at dissections and conducted post-mortems. Through Guy de Chauliac the tradition of Mondino passed to Montpellier.

Early in the next century Pietro d'Argellata, a Bolognese professor, examined the body of Pope Alexander V, who died suddenly at Bologna in 1410. Pietro made a post-mortem examination of the body of the pontiff, and included an interesting description of it in his work on *Surgery*. This is, however, barely Anatomy, and anatomical advance was at a standstill until nearly the end of the fifteenth century.

Men who made a great impression on their own age were Gabriele de Gerbi (died 1505) and Alessandro Achillini (1463–1512). De Gerbi, educated at Pavia, and professor at Verona, was a verbose and tiresome writer whose evil influence may have done something to delay anatomical advance. His work, largely taken from Mondino, bears the authentic scholastic stamp. It is claimed for him that he distinguished the olfactory nerves, and that he paid much attention to development, on which, however, he wrote a work of little worth. He introduced into Anatomy the term *Pilorium*, later purified into *Pylorus* by Vesalius. More noticeable was Alessandro Achillini, who divided his activities between the schools of Padua and Bologna. He was an extremely disputatious and windy controversialist who wrote verbosely on Philosophy. His anatomical work, like that of De Gerbi, is largely a commentary on Mondino. Nevertheless, his writings contain a

few additions to anatomical knowledge, and it is clear that
Anatomy is at last again astir. Achillini described Wharton's
duct a century and a half before the man whose name is
associated with it.   He redescribed the infundibulum,

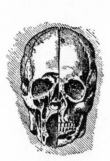

FIG. 43.

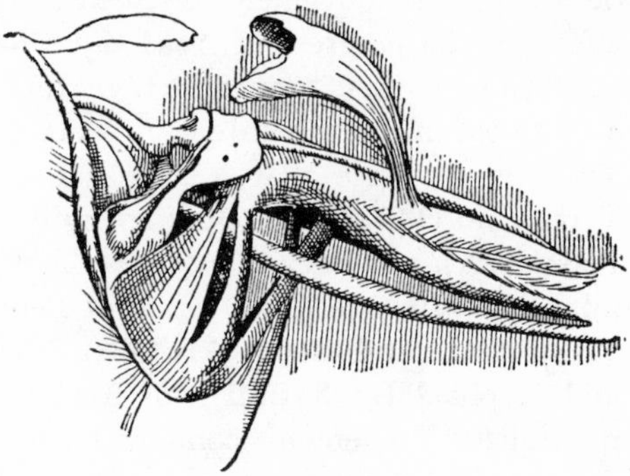

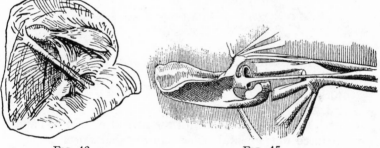

FIG. 44.

FIG. 46.   FIG. 45.

ANATOMICAL SKETCHES OF LEONARDO.

FIG. 43.—Dissection of the skull showing maxillary antrum, frontal
    sinus, and nasal fossæ (Fogli B, folio 41 verso)—after Holl.
FIGS. 44 and 45.—Dissection of the muscles of the shoulder (Fogli A,
    folio 2)—after Holl.
FIG. 46.—The right ventricle laid open showing the tricuspid valve
    and the intra ventricular moderator band (Quaderni iv, folio 13).

observed the inferior cornu of the anterior ventricles and rediscovered the fornix. His description of the cæcum was an
advance on that of Mondino, and he improved also on the
current descriptions of the duodenum, ileum, and colon.
It is often said that Achillini described the two auditory
ossicles, Malleus and Incus, but this seems to be an error.

§ 8   *Naturalism in Art, about* 1450–1550

Leonardo da Vinci

In the fifteenth century there made itself felt in the realm
of Art a great movement destined to react with far-reaching
effect on the progress of Anatomy. *Naturalism*, born in the
thirteenth century, now came to maturity. Artists were taking
an interest in the accurate representation of the human form.
There is evidence that a long line beginning with Andrea
Verrocchio (1435–88), including Andrea Mantegna (died 1506,
Plate XIX), and Luca Signorelli (about 1444 ?–1524), and cul-
minating with the giant forms of Leonardo da Vinci (1452–
1519), Albrecht Dürer (1471–1528), Michelangelo (1475–1564),
and Raphael (1483–1521), all used the scalpel. The great
naturalistic movement combined with the improved access

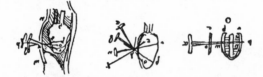

FIG. 47.—An experiment by Leonardo on the heart. Needles are
thrust through the chest wall of the pig into the substance of the
heart and the movement of the organ can thus be followed
(Quaderni i, folio 6).

to Greek sources that came with the revival of learning, and
so produced fundamental changes in the anatomical outlook
which found their most natural and forceful expression in
Vesalius.

Of the practical anatomical knowledge of Michelangelo,
Dürer, and Raphael, we have ample evidence. All have left
drawings of dissections. One of the very greatest anatomists
was, and is, Leonardo da Vinci (Plate XIII). That marvellous
man doubtless began to dissect to improve his Art. Soon, however, he became interested in the structure and workings of the
body. His scientific pre-occupation at last exceeded his
artistic, and his anatomical notebooks, published in recent
years, have revealed him for what he was, one of the very
greatest biological investigators of all time. In endless matters
he was centuries ahead of his contemporaries. Had he pro-

PLATE XI

DRAWINGS OF HEART BY LEONARDO

A.   The Figure to the left is from Quaderni II, folio 3 verso and has been lettered according to modern notation.
B.   The Figure to the right is from Quaderni I, folio 3.   It is a diagram showing the structure of the heart and exhibiting the
      passages in the septum hypothecated by the Galenic physiology.   It also shows Leonardo's looking-glass writing.

duced the anatomical textbook which he had planned in collaboration with the Pavian professor, Marcantonio della Torre (1481–1512), the progress of Anatomy and Physiology would have been advanced by centuries. The early death of della Torre prevented this, and Leonardo's anatomical manuscripts remained hidden till our day. Yet Leonardo occupies so isolated a position that it would destroy our perspective if we dwelt long upon him. He cannot, in fact, be properly considered in the series of anatomical discoveries, but must be taken by himself.

In Osteology, Leonardo is the first to draw adequate figures of the skeleton and to adopt the modern method of repre-

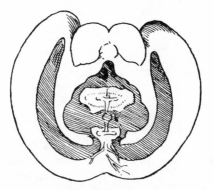

FIG. 48.—Wax cast of ventricles of the brain as portrayed by Leonardo (Quaderni v, folio 7).

senting the parts from front, back, and side. He has first-rate figures of the maxillary antrum and of the frontal sinus (Fig. 43). He recognizes the different types of vertebræ perfectly, and catches exactly the curves of the spinal column.

In Myology, Leonardo excelled. As an artist he had specially studied the surface muscles ; such errors as he there makes are often wilful, the result of his own rich and strange imagination. He has remarkable diagrams exhibiting the action of the muscles. His figures of the diaphragm are truly wonderful. His drawings of the structure of the hand are well-nigh perfect. He has particularly clear figures of the shoulder muscles (Figs. 44 and 45).

In Angeiology some of Leonardo's figures illustrate the
physiology of Galen (Plate XIB), others are marvels of observation and insight. He has admirable illustrations exhibiting the
distribution of blood vessels in leg and arm. Best of all are
certain of his drawings of the heart (Plate XIA), some of
which exhibit the intra-ventricular moderator band (Fig. 46)
centuries before it was recognized by anatomists. He made
experiments on the movements of the heart repeated by
Harvey (Fig. 47). He also constructed models to illustrate
the action of the valves.

Among Leonardo's triumphs in neurology are some extraordinary figures of the brain. He succeeded in injecting the
ventricles with a solidifying medium—itself a difficult

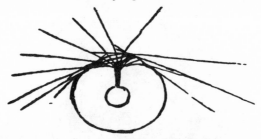

FIG. 49.—Diagram of eye by Leonardo showing the *sphæra crystallina*,
the supposed seat of vision, in the centre of the globe (Quaderni iv,
folio 12 verso).

operation — and obtained casts of the cavities these are
the first known attempts at anatomical injection (Fig. 48).
His ideas on the structure of the eye are little in advance
of his time (Fig. 49).

Leonardo paid much attention to the generative organs
(Fig. 50). Some of his figures of those parts, notably that of the
child in its mother's womb, are real masterpieces. He had a
good idea of the mechanism of parturition. Curiously he
figured the placenta as cotyledonous.

It has been mooted whether Leonardo's figures may not have
affected Vesalius. So far as direct influence goes, the answer
is in the negative. But the atmosphere created by Leonardo
and the other great artist anatomists did certainly bear fruit.
In that sense the work of Leonardo was not wholly lost, and
there are even instances in which the actual mode of repre-

PLATE XII

EARLY DIAGRAMS OF EYE

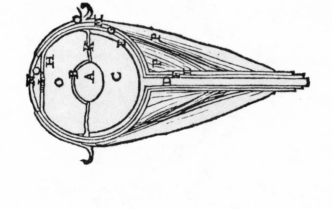

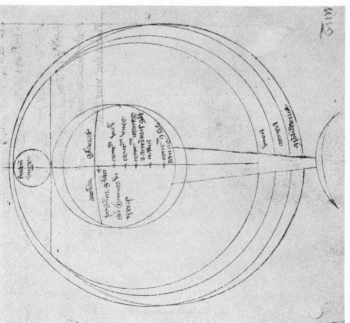

A. From a Thirteenth Century MS. of Roger Bacon at the British Museum (Roy. 7, F.VIII, folio 50 verso). In front is the foramen uvee (pupil). Occupying the centre of the globe is a spherical "crystalline lens" divided into two parts, an *anterior glacialis* and a *posterior glacialis seu humor vitreus*. The structure of the eye is developed along mathematical lines.

B. From the first edition of the *Fabrica* of Vesalius (1543). The crystalline humour (A) still occupies the central position. No great advance has yet been made in ophthalmic anatomy.

[*face p.* 92

sentation adopted by Vesalius bears some resemblance to
that of Leonardo.   The naturalistic movement in Art which
Leonardo represented had, however, the profoundest influence
in Anatomy.   Without it the subsequent work of Vesalius
would have been impossible.

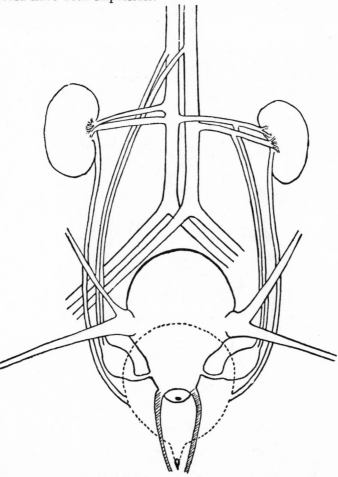

FIG. 50.—Tracing from Leonardo of the outline of the urino-genital
system (Quaderni i, folios 11 and 12).   It contains many errors.
Among them the uterus is given *cornua*. (Compare figures 39,
51, 62.)   These are of peculiar form and proceed from the body
of the organ. (Compare Fig. 51, in which they are of similar
form.)   It is probably an early anatomical conception of Leonardo.

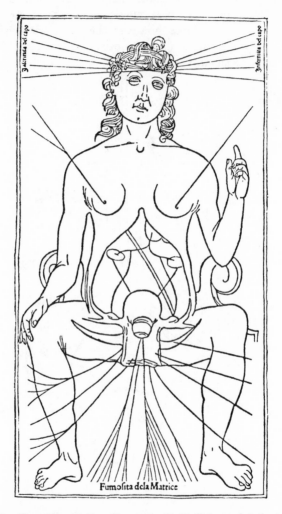

Fig. 51.—From the Italian *Fasciculo di Medicina* printed at Venice in 1493. The anatomy though very erroneous exhibits the uterus naturalistically treated. This is the first figure in a printed book in which an internal organ has been drawn from the object. Some have thought the figure was drawn or at least influenced by Leonardo ; note the curious lateral projections from the body of the uterus and compare them with those in Fig. 50.

PLATE XIII

LEONARDO DA VINCI 1452-1518
by himself, from a pastel in the Royal Library,
Turin.

§ 9   *The First Anatomies Printed with Figures*, 1490–1545

Printing with moveable type, invented about 1450, hardly came into extensive use for medical purposes until the last decade of the century.   The atmosphere of Art was not slow in reacting on this professional literature.   The first instance which can be clearly traced is in an Italian work printed at Venice in 1493, under the title of *Fasciculo di Medicina*.   It is

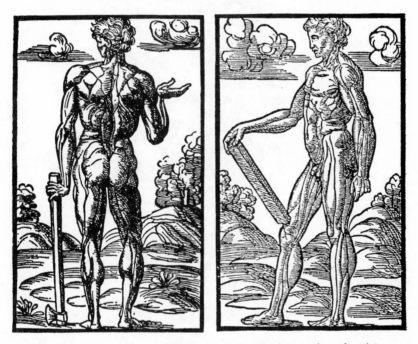

Fig. 52.—The surface muscles drawn for the instruction of artists, from Berengar of Carpi *Commentaria* on Mondino—Bologna, 1521.

a collection of tracts and bears on its first page a portrait of Pietro da Montagnana (died about 1460), a professor at Padua who claimed to have witnessed a number of post-mortem examinations.   The longest text is the *Anothomia* of Mondino, illustrated by a magnificent woodcut of a dissection scene, perhaps our best representation of an Academic " Anatomy " (Fig. 35).   Another item is a fine figure of the female anatomy

in which the uterus is drawn from the object (Fig. 51). This
is the earliest instance in a printed book of a sketch of an

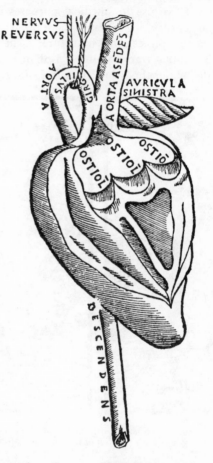

FIG. 53.—The heart from Berengar of Carpi *Isagogæ breves*, Bologna,
    1523. The left ventricle is opened and there are shown the
    ostiolæ or valves at the root of the aorta. There is also seen the
    left *nervus reversus* or recurrent laryngeal nerve and a spiral
    auricula sinistra.

organ of the body. It is from the hand of an excellent
draughtsman. A resemblance to certain of the figures of
Leonardo has been pointed out.

The artistic spirit thus fermenting in the medical schools soon bore further fruit. From this time on we begin to have illustrated anatomical textbooks, of which the first is that of Jacob Berengar of Carpi (died 1550). He was professor of Surgery at Bologna from 1502 to 1527. During that period he performed many dissections, and published two important anatomical works. One, though modestly put forward as a commentary on Mondino, is in reality an original contribution of considerable value. It is the earliest anatomical treatise that can properly be described as having figures illustrating the text. These figures vary in excellence. Some are not devoid of beauty, and are prepared for the use of artists rather than anatomists (Fig. 52). Carpi does not hesitate to criticize the work on which he professes to comment. Thus, for instance, he denies the existence of the *rete mirabile* below the brain, thereby contradicting not only Mondino but also Galen.

Carpi is the first to describe the vermiform appendix, the first to see the arytenoids as separate cartilages, the first to recognize the larger proportional size of the chest in the male and of the pelvis in the female, the first to give a clear account of the thymus gland. He knows something of the action of the cardiac valves (Fig. 53). His description of the brain is an advance on Mondino, recognizing the general form of the ventricles, the composition of the choroid plexus out of the arteries and veins, the pineal gland and the relations of the fourth ventricle. The language of Berengar is excessively debased, and among the writers we have mentioned he is perhaps the worst Latinist—and this is a position of no mean distinction! Probably his language prevented his works from being more widely read. He is responsible for the barbarous term *Vas deferens*. A contemporary term, derived from no language at all, is *Synovia*, invented by the alchemist, quack, rebel, prophet, and genius, Philip Aureolus Theophrastus Bombastus von Hohenheim, called Paracelsus (1493–1541), whose work lies outside our sphere.

A peculiar development of Anatomy in the first half of the sixteenth century was the vogue of the so-called " fugitive anatomical sheets ". On these were printed figures showing the bare outlines of Anatomy. They were largely used by students of Medicine who then, as now, had no desire to burden their memories with superfluous details. These fugitive sheets are usually inferior to the illustrations in the anatomical textbooks. There is a group of fugitive sheets, however, which are more important for the history of Anatomy. These

FIG. 54.—Johannes Dryander, of Marburg.

were intended not for medical, but for art students, and show the continuing interest in Anatomy in the quarter from which the real reform of Anatomy had come.

An illustrated edition of Mondino's Anatomy was issued by Johannes Dryander, of Marburg (died 1560, Fig. 54), in 1541. It contains a good many figures. Some are quite good, and there is evidence that these were stolen from Vesalius, who was then at work preparing his masterpiece (Figs. 55, 56). They are peculiarly interesting as exhibiting an otherwise unrepresented stage in the development of the great Reformer of Anatomy.

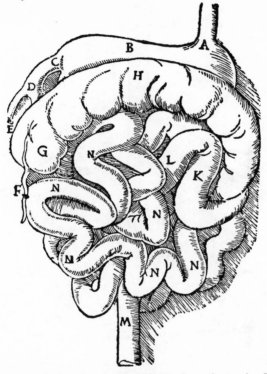

FIG. 55.—Figure of viscera from Dryander, *Anatomia Mundini*, Marburg, 1541. The legends are thus explained in the original: " This figure shows the position of the intestines. A is upon the fundus of the stomach into the interior of which food and drink pass when it is relaxed ; B is the middle part of the stomach ; C is the lower part of the stomach called *portanarius* ; DE is the passage from the gall bladder by which bile is often poured into the stomach ; F is the additamentum of the large intestine in the cæcum ; G indicates the cæcum (*monoculus*) ; KL colon ; M rectum ; NNN the remaining small intestines." This is the earliest figure showing the vermiform appendix. Compare, however, the figure from Vesalius, Fig. 68, *quarta*. Dryander probably took his drawing from an early sketch of Vesalius.

The most fully illustrated of the pre-Vesalian Anatomies is that of the Frenchman Charles Estienne (1503–64) who sprang

of an eminent family of humanistic printers. He did not
dissect publicly. Part of his work, which appeared in 1545,
was in preparation as early as 1530, and is thus less than a
decade later than that of Berengar, and more than a decade

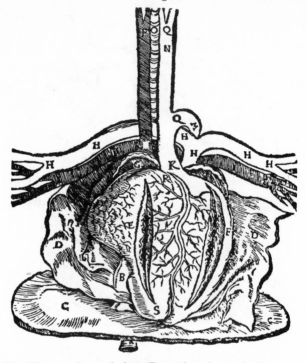

Fig. 56.—Heart and vessels from Dryander, *Anatomiæ hoc est corporis
humani dissectionis*, Marburg, 1537.   The same figure is to be
found in Dryander's *Anatomia Mundini*, Marburg, 1541.   The
legends may be thus translated :—

" CC diaphragm ; DBD involucrum of heart, a thin membrane
surrounding heart and filled by its substance ; E right ventricle ;
F left ventricle ; G right auricle ; HHHH course of *arteria venalis*
(pulmonary vein) ;   III course of *vena arterialis* (pulmonary
artery) ;   KQ beginning of aorta ;   QM aorta ascendens ;
QN aorta descendens ;   O trachea ;   P great ascending vein ;
R veins nourishing heart ; S extremity of the heart."

earlier than that of Vesalius. His hideous figures are copies
of those of contemporary artists with anatomical details added
(Fig. 57). These illustrations are, however, the earliest, except
those of Leonardo, in which whole systems, venous arterial, or

nervous, are shown. The text is largely dependent on Galen. Estienne's best department is, perhaps, that of arthrology, and he has good descriptions of the clavicular joints, of the

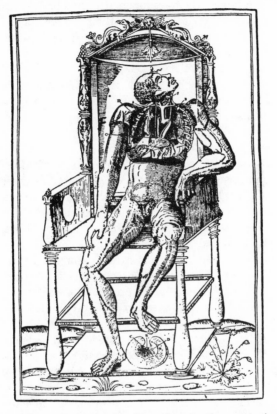

FIG. 57.—Charles Estienne, *De dissectione partium corporis humani*, Paris, 1545. The thorax is opened and the heart and lung removed. The arch of the aorta is seen with the vagus nerve A and B on either side and the recurrent laryngeal on the left. The pleura is shown turned back at D and the œsophagus at C penetrating the diaphragm E. At the foot is a very poor figure of the diaphragm, showing the membranous part at A, the foramen for the œsophagus at B, that for the aorta at C, and for the vena cava at D.

temporo-maxilliary articulation and of the joints and ligaments of the spine. He was the first to trace blood-vessels into the substance of bone. His figures display the vermiform appendix, though in this he was preceded by Dryander (1541). Estienne does not refer to the appendix in his text. He was the first to remark upon the valves in the veins, though of their function he had no inkling. He gives much attention to the form of the muscles, drawn separated from their attachments. Most remarkable of his observations is that of the canal in the spinal cord, which was not again remarked upon until the work of J. B. Senac (1724).

Estienne seems to have had little difficulty in obtaining material for dissection, though he discusses whether the bodies of apes or men should be used. He claims to have seen with his own eye all that he describes. For this, however, he must have had the eye of faith, for he describes structures found in the text of Galen, but not in the human body. He lays much emphasis on glands of which he describes the parotid, the thymus, the lymphatic glands at the roots of the mesentery, and in the armpit and groin, and apparently the lachrymal glands. He injected the blood-vessels with air. He gives an extremely bad and very Galenic figure and description of the vascular system. He has a fairly correct figure of the spleen. The book is one of our best sources for estimating the state of Anatomy immediately preceding Vesalius.

The last of the pre-Vesalian illustrated anatomies that we need consider is that of Giambattistta Canano of Ferrara (1515–79). He was long professor of Medicine in his native town. About the year 1541 he produced a small pamphlet describing the muscles of the arm. He intended to follow it up by others dealing with other parts of the body. The tract was unique for its time in exhibiting each muscle in a separate figure and in approximately correct relation with the bones (Plate XIV). The beautifully drawn figures, owing to the defective paper, have greatly deteriorated since his day. The descriptions are succinct and apt. Had Canano continued the publication it would have made a real landmark in the history of Anatomy. The Jewish physician Amatus Lusitanus (1511–62 ?) dwelt in his house, and with him investigated the valves of the veins. Perhaps discouraged by the appearance of the great work of Vesalius, Canano unfortunately ceased from publication. The future lay with Vesalius.

Plate XIV

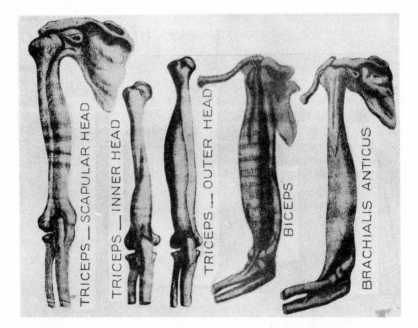

ARM MUSCLES FROM CANANO

The work of Canano *Musculorum humani corporis picturata dissectio* appeared at Ferrara, probably in 1541. Only the first part appeared and only about twelve copies of the work have survived. It is illustrated with copper plates now much deteriorated. In all the copies the print has come through into the figures and the paper is foxed. This plate is in the nature of a " restoration " It has been obtained by photographing the Dresden copy of the figures. The photographs were then carefully cut away from their background and stuck on a white sheet of paper. Professor Sudhoff of Leipzig kindly supplied the photographs.

§ 10   *The Humanists*, 1450–1550

We have so far considered only practical anatomists.
Toward the end of the fifteenth century there arose a class of
medical scholar, more concerned with writing than dissecting,
but whose influence in the course of Anatomy was so profound
that it can hardly be passed over. The fifteenth century saw
the beginning of a revolution in education. That revolution,
based on the recovery of the ancient classics, had its effects in
the study of medicine, notably in Anatomy. We have seen
that in the Middle Ages the current anatomical treatises
were almost entirely translated from Arabic. Mondino in the
early part of the fourteenth century was just beginning to
gain access to translation direct from Greek. During the two
centuries since Mondino, there had been a progressive increase
in these translations from the Greek. They represented not
only the ancient writers better than the old-fashioned Latino-
Arabic translations, but they also represented better the
anatomical facts.

This movement in Medicine was only a part of a great
intellectual movement to which the name *humanist* is
attached. Humanism affected every department of mental
activity. A great number of the humanists were physicians,
and not a few were interested in Anatomy. By the middle of
the sixteenth century these men had recovered practically
all the medical classics that we now possess. Naturally
their knowledge of these ancient writings was less well arranged
than ours, and critical study was far less advanced than with
us. It is, however, substantially true to say that they knew
as much as we do of the important anatomical works
associated with the names of Galen, Hippocrates, and Aristotle.
Complete Greek texts of these and of other ancient medical
authors became available in good editions in the first quarter
of the sixteenth century. All were early translated into good
Latin versions. Moreover, the knowledge of the Latin language
was itself improved by the recovery and publication of the
Latin classics.

In the course of the humanistic revival of Medicine,
the first published work of real influence was an edition

of Celsus, which appeared in 1478 at Florence. This
made a considerable change in anatomical knowledge, for
Celsus had been quite unknown in the Middle Ages. His
excellent Latin created a new standard for medical writing,
and many of his anatomical terms came into use.
These replaced Arabic and Latino-Arabic words. Some
of these words taken from Celsus are among the commonest in our anatomical nomenclature, and have remained
in use to this day; among them we note *Abdomen, Anus,
Cartilage, Humerus, Occiput, Patella, Radius, Scrotum, Tibia,
Tonsil, Uterus, and Vertebra.*

The preparation of translations of Galen occupied a whole
host of learned and able medical authors. The corpus of
Galenic works provided anatomists with a source of new and
exact terms and these again replaced many of the old Arabic
and Latino-Arabic words. From these Latin translations of
Galen, printed during the sixteenth century, there entered
Anatomy a very large number of terms; among them
*Allantois, Anastomosis, Aponeurosis, Apophysis, Arytenoid,
Azygos, Carotid, Choroid, Condyle, Cremaster, Epididymus,
Ginglymus, Glottis, Gomphosis, Hyaloid, Masseter, Meconium,
Olecranon, Pancreas, Peritoneum, Psoas, Thyroid, Torcular,
Ureter, Zygoma.*

Alessandro Benedetti (about 1455–1525) took a large part
in opening the " humanistic period " of Anatomy. He marks
also the rise of Padua as a centre of anatomical study. He
studied at Padua and afterwards spent a long time in Greece,
where he learnt the language. Returning to Padua, he founded
there an anatomical theatre where he demonstrated to very
large audiences. In 1493 he published *Five books of Anatomy,
on the history of the human body.* It contains no new facts. It
appeals, however, direct to the Greek Galenic texts passing
lightly over the Latino-Arabic versions. We owe to him our
term *valve* applied to structures in the heart. His contemporary
and colleague, Antonio Beniveni (about 1450–1502), left a series
of records of post-mortem examinations which were published
posthumously in 1506. This is the first work on Morbid
Anatomy.

Of the many humanists who occupied themselves with the

actual work of translation one is of special interest to English readers. Thomas Linacre (1460 ?–1524), physician to Henry VIII, tutor to the Princess Mary, founder and first president of the College of Physicians, a benefactor of both the ancient Universities, and one of the earliest, ablest, and most typical of the English Humanists, spent much energy in this work, for which his abilities peculiarly fitted him. He studied Greek at Padua, which had become a humanist centre. On his return to England he translated no less than six important works of Galen, most of which had anatomical bearing. He added nothing to anatomical knowledge by direct observation, and he looked rather to the form than the substance of the works of the ancient writers.

There was another class of medical humanists, who were occupied in forming synopses, abstracts, and summaries in their own words of the ideas of the ancients. The most typical of these was another Paduan student, J. B. Montanus (1498–1551). He spent much of his energy in expounding the anatomical and physiological views of Galen. It was largely through his influence that the Arabs passed into the shadow in the North Italian Universities. Another member of the same group who stamped himself very deeply on his time was Johannes Günther, of Andernach (1487–1574, Fig. 58). He is, moreover, interesting as representing along with Linacre the spread of Humanism beyond Italy to the North-West of Europe.

Günther exerted influence less by his writings than through his pupils, to whom he endeared himself. He taught at Paris, where he had as students Vesalius (p. 112), Servetus (pp. 113, 140), Rondelet (p. 147), and Dryander (p. 98). Günther was a fine Greek scholar, and translated into Latin many works, including the great treatise *On anatomical procedure* of Galen. More interesting for us are his *Anatomical institutions according to Galen* (1536) and his *Medical knowledge and practice in ancient and modern times* (1571). The first was later edited by Vesalius, and gives, along with the anatomical work of Estienne (p. 100), the best survey of the humanistic pre-Vesalian Anatomy. The second, published after the death of Vesalius, shows the influence of the new Anatomy in an author of the old school who was by then very advanced in years and had watched the rise of Vesalius. Regarded as a practical anatomist, Günther's achievements are negligible. There

FIG. 58.—Johannes Günther, of Andernach (1487–1574).

can be no doubt, however, that he had occasionally dissected, and that he did something to establish a tolerable anatomical nomenclature.

In connexion with the process of " purification " of anatomical nomenclature and the substitution of Greek for Arabic terms an important agent was an otherwise unimportant ancient writer Julius Pollux (A.D. 134–92). He was a contemporary of Galen. His work *Onomasticon*, without contemporary importance and unknown in the Middle Ages, was quite without influence until printed in 1502. We have therefore refrained from considering it earlier. The *Onomasticon*—the word means simply " vocabulary "—was dedicated to the Emperor Commodus, the son and heir of Marcus Aurelius, and the patient of Galen (see p. 49). It consists of a series of sections, each containing a list of the most important words relating to a particular subject. Attached to these words are short explanations, often with quotations from ancient writers. One of the sections treats of anatomical terms. The text thus published in the sixteenth century became a sort of storehouse from which the Humanist physicians drew words to replace the Arabist terms in current use. Pollux has thus become the source of a good deal of our modern anatomical terminology. From him there have entered Anatomy the words *Amnion, Antihelix, Antitragus, Atlas, Axis, Canthus, Clitoris, Cricoid, Epistropheus, Gastrocnemius, Tragus,* and *Trochanter.*

At the end of the fifteenth century practical Anatomy came to be recognized in the University of Paris. The first eminent anatomist of the Paris school was Sylvius (Jacques Dubois, 1478–1555, Fig. 59), whose unlovely character has robbed him of much of the distinction that would otherwise be his. After having pursued other studies, he took a medical degree at Montpellier at the age of 51. He came to Paris, where, in 1531, he began to teach before crowded audiences. There must have been something in his mode of address which gave him so wide an appeal. That something we can perceive in his varied learning and in his admiration for and knowledge of Galen. He is a Humanist getting into touch with practical Anatomy. Obloquy has since fallen on his name in connexion with his unfortunate relations with Vesalius. Yet there can be no doubt that Sylvius was a very able exponent and a man with great capacity for systematic statement. He practised

the art of injection, which though perfectly well known to
Leonardo da Vinci and Estienne, seems not to have been
used in academic Anatomy before his time.

Looking back on the Paris school from the vantage
ground of a later time, we are liable to see it through
the eyes of Vesalius. The Reformer of Anatomy, like
his teacher, Sylvius, was a child of his age, and it was a

FIG. 59.—Jacobus Sylvius from a contemporary print.

foul-mouthed age. Men habitually spoke of their colleagues
in a way that we should now regard as not only unseemly
but indecent. Sylvius did not deserve all the ill that has
been spoken of him. Despite his Galenism, he made real
additions to knowledge. Thus he described the sphenoid
for the first time, and the jaw-bones and the vertebral column
far better than any previous writer. His power of

systematization enabled him to make improvements in nomenclature, introducing for instance the term *Corpus callosum.* (It is not his name, however, but that of another Sylvius who lived in the following century that has been given to the *Fissure of Sylvius.*) Anatomical nomenclature was in great disorder till the time of our Sylvius, and there can be no doubt that in this, as in other matters, Vesalius owed him a not inconsiderable debt.

Sylvius, living in France and away from the great Italian revival of Art, was quite uninfluenced by that great movement. He poured scorn on the new-fangled use of figures for the illustration of anatomical facts. In this he presented a sharp contrast to his pupil. But even Vesalius, until he reached Italy and came under the spell of Renaissance Art, was but an able, energetic, and learned young Galenist, not sharply differentiated from the older school to which his master belonged. It was the incentive provided by Art which completed his equipment. To the achievements of the great Reformer of Anatomy we now turn.

Fig. 60.—Vesalius (1514–64) from the *Fabrica*, 1543.

IV

MODERN TIMES TO HARVEY, 1543–1628

§ 1 *Vesalius, the Reformer of Anatomy* (1514–64)

FEW disciplines are more surely based on the work of one man than is Anatomy on Vesalius. And yet it can be said that he is, in a sense, a lucky man in the position that he holds in the scientific world. His great work was not the result of a long lifetime of experience, as was that of Morgagni or of Virchow ; it was not wrought in the fierce heat of an intellectual furnace as was that of Pasteur or of Claude Bernard ; it was not a task of subtle reasoning and skilled experimenting, as was that of Harvey and of Hales. Vesalius was a very characteristic product of his age. The womb of Time was in Labour, and it brought him forth. His intellectual father was the Galenic Science that had gone before him. His mother was that fair creature, the new Art, then in the very bloom of her youth. Until these two had come together there could be no Vesalius. When these two had come together there had to be a Vesalius. If it be genius to be such a product of one's age, then Vesalius was a genius. He was a strong resolute man of clear, firm-knit, and unsubtle mind, and he fulfilled that for which his father and his mother had begotten him. He did no more, and he did no less. If, unrevealed by his one great work, the rock of his soul held yet further fastnesses, he so concealed them that no man has since entered therein, despite all the searchings of the scholars.

Andreas Vesalius was born of a medical family at Brussels. As a boy he was constantly dissecting the bodies of animals. Such tastes are common enough with boys nowadays. Nature study is taught in our schools, and it is easy to get help and hints from books. In the days of Vesalius the study of Nature was little regarded, and no such books had been written. Vesalius had to find his own way.

He studied first at Louvain, afterwards at Paris under
Sylvius and Günther. He was thus subjected to a very full
and complete Galenic training. We have seen how Anatomy
was taught in mediæval times, and the methods of instruction
at Louvain and Paris had not greatly improved, though the

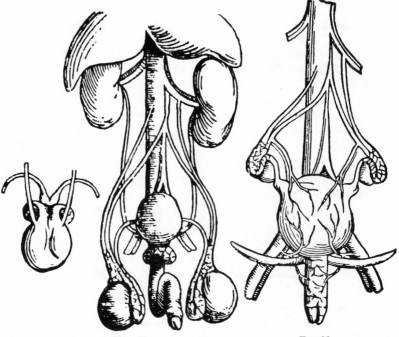

FIG. 61.   FIG 62.

FIG. 61.—The male genital system from the *Tabulæ sex* of Vesalius.
Venice, 1538.

FIG. 62.—The female genital system from the *Tabulæ sex*, 1538.   The two
figures and notably Fig. 62 exhibit many traditional errors.   Thus the
uterus has, between neck and body, the two cornua of mediæval tradition
and the cervix is confused with vagina.   An endeavour is made, however,
to establish an homology between the structures in the two sexes.

texts were vastly better and had undergone revision by the
Humanists.   With Sylvius, Vesalius quarrelled.   Between
Günther, a learned and amiable man, and the young Vesalius
there subsisted an affectionate mutual regard which the
witty acidity of the irrepressible tongue of Vesalius did not

wholly destroy. Another pupil of whom the aged Günther also spoke kindly was that tragically doomed Servetus (Plate XV), as great a contrast to Vesalius as one man may be to another.

The first published anatomical work of Vesalius was a revision of Günther's *Anatomical Institutions according to Galen* (Venice, 1538). His next work appeared in the same year,

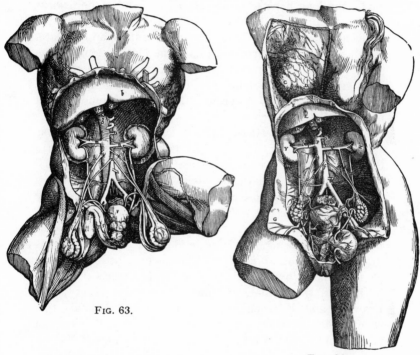

FIG. 63.

FIG. 64

FIG. 63.—Male urinogenital system from the *Fabrica* of Vesalius, 1543.
FIG. 64.—Female urinogenital system from the *Fabrica* of Vesalius, 1543.

These figures should be compared to those from the *Tabulæ sex* of 1538, figs. 61 and 62.

and was a set of six fugitive sheets to illustrate the Galenic *Anatomical Institutions*. These *Tabulæ sex* (Figs. 61, 62) were printed after Vesalius had left Paris and had come to Italy, whither, without ado, we may follow him. Great historical interest attaches to these first two anatomical works of the future Reformer of Anatomy, but they are not in any way revolutionary. The text is frank Galenic Anatomy. True, his fugitive sheets are on a more generous page than any of their predecessors. They are better drawn and more detailed. But they still show the five lobed liver, the *rete mirabile*, the venous system arising from the liver, the portal vessel conveying chyle, the long hepatic vein. They also exhibit the *truncus brachiocephalicus* of the dog, the long, protruding coccyx, of the ape, and the sternum in seven segments of an ungulate. All these are characteristic traits of the old Anatomy. The plates are in fact Galen, well and diagrammatically portrayed. They give a clearer visualization of the working and structure of the body than had yet been set forth, but no fundamental change. Vesalius was not yet 24 years of age.

At Padua he settled, and at the end of 1537 was appointed professor at the University. His basic reform there was to do away with " demonstrators " and " ostensors " in the old sense, and to put his own hand to the business of dissection (Frontispiece). Such a change he had already sought to make, but found difficult in the always conservative environment of the University of Paris. Vesalius was soon lecturing to large audiences, as the famous frontispiece to his work sets forth. His demonstrations were on the human body, and he used living models on whom he marked the outlines of the joints and of other parts. Drawings and skeletons were always at hand. Animals, too, were available for experiment and dissection, and there can be no doubt that, not infrequently, he availed himself of them rather than of the human subject for anatomical description. He was now a Professor on his own account, master in his own department, stimulated by the applause of a concourse of students. The work was carried on with great energy and drive. Every line and every figure of his great book, the product of but five years' activity, is instinct with his virile power. That splendid monograph *On the fabric of the human body* was completed in 1543. Vesalius was still but 28. It was issued at Basel almost simultaneously with a companion *Epitome*, which, however, bears the date of the previous year With these two books, the life-work of Vesalius was completed.

The second edition of the *Fabrica* appeared in 1555, and contains some improvements, but no fundamental changes. From 1543 onward Anatomy becomes Vesalian, while Vesalius himself passes into the background. His life course is well known and is easily accessible. We shall not follow it for it is not of primary importance for the history of Anatomy. For that purpose Vesalius and his great work *On the fabric of the human body* are one. Without the book he would be but a ghost.

§ 2   *Threefold Character of Vesalius : Artist, Humanist, Naturalist*

It may be pointed out that Vesalius has not given his name to any part of the body. In this he differs from many anatomists. We have the canal of Eustachius, the tube of Fallopius, the duct of Botallus, the circle of Willis, the lobe of Spigelius, the fissure of Sylvius, the glands of Bartholin, the island of Reil, the ganglion of Gasser, the cartilage of Arantius, the sinus of Valsalva, the tubercle of Lower, the valves of Morgagni, even the torcular of Herophilus and the veins of Galen. As for Vesalius, he has left his name on the whole fabric of the human body. It was indeed the body as a whole that Vesalius had always in his mind's eye—in the background of his figures, as it were. It was this vision that in making him a great creative anatomist made him, at the same time, a great creative artist. And yet it is just this element in him that makes his work such very difficult reading for us moderns. To understand Vesalius there are certain ideas that come to us almost instinctively of which we must rid ourselves. To understand him we must try to think like Renaissance artists and not like modern evolutionists.

The modern scientific anatomist deals primarily with description and secondarily with origin. He first takes an organ or a part for its own sake and investigates it in detail. Then he treats it comparatively and embryologically, but always from an evolutionary point of view. Not so Vesalius. For him the body is a fabric, a piece of workmanship by the Great Craftsman. The parts he is examining must be fitted into the system as whole. Furthermore, his Anatomy is essentially *living* Anatomy. The parts are not the subject of morphological treatment, but of investigation as contributing to the existence of that complex vital unit we call a Man. Thus his great general figures of muscles or of bones are not placed in the diagrammatic positions to which we are accustomed in our textbooks, where the parts are shown from front, from back, or from side. By him they are posed as in the living body, and given a background such as that to which they were accustomed during life. The method is used by Vesalius with all the artist's skill, and this was his own. Yet the technique was of his age. We find the same method employed, though more clumsily, by Berengar, by Dryander, and by Estienne. It will be observed that we do not discuss by whose hand the actual drawings were made. For our present purpose this is strictly irrelevant since the artist's mind that conceived them was surely that of Vesalius himself.

Moreover, and this perhaps is yet more important, Vesalius, as a child of his age, could not even as an anatomist help thinking of the end for which man was made. Remember he was steeped in Galen, and it is too much to ask that even he should wholly shake off the yoke of Galenic teleology. But with an artist's mind and eye, Vesalius transmuted that age-old moss-grown scheme into something higher, nobler, more worthy of labour. For him Man is a work of art, God is an artist. He was no philosopher, nor must we seek in his pages for any formal justification of this view. But so much he says, and says well, over and over again. Men and women he saw, as it were, as the Artist's " studies " for his great design. Imperfect studies, indeed. Vesalius did not, like Galen, harp constantly on the perfection of man's form. He had, as we know, criminals, worn-out paupers and bodies wasted by disease on which to practise his art ; yet, such as they were, worthy of our attention as showing forth, however distantly, the design of Man in the mind of the Godhead. To reach closer than these poor corpses to that grand design was the real aim of the anatomist. We think of Anatomy in terms of Evolution, and our question is always " whence " ? and " how ? " Vesalius

thought of Anatomy in terms of Design, and his questions,
had he been philosophically articulate, would have been
" whither ? " and " why ? "

The vigorous teeming mind of Vesalius presented yet
another aspect on which too little stress has been laid. He
was a very learned man, one well acquainted with all the new-
found wealth of Antiquity which the Humanists were
making more and more accessible to the reading public. He
himself, in the manner of his time, was not a little vain of his
erudition. He makes a great and, to modern eyes, an unnecessary display of his learning in the *Fabrica* and even more
in the *Epitome* and in the *Tabulæ sex*. The books are full of
Arabic and Hebrew words. To these he was helped by other
scholars. Of those languages his knowledge was, in fact, of
the slightest. In judging him we must always remember the
age in which he lived. The great intellectual battle was
still being waged between the " Arabists " and the
" Humanists ". The Humanists were winning, but as yet the
issue of the conflict was by no means clear. Thus the constant
reference to Arabic and Hebrew was something more than mere
vain show, though it must be admitted there is, too, something
of the showman in Vesalius. With the Greek language, however, he had considerable facility. So far as the Latin
translations of the Greek, Arabic, and Hebrew classics were
concerned, there can be no doubt that Vesalius was a highly
accomplished man and more critical than most scholars of
his time.

Vesalius occupied himself to a considerable extent with
editing the actual works of Galen. Many terms of Galenic
origin thus naturally entered his vocabulary. Many more
he borrowed from Günther and the other humanist physicians,
and from the editions of Celsus and Pollux. Some anatomical
terms he invented for himself. To him, for instance, we owe
Atlas, applied to the first cervical vertebra (Pollux who uses
the word applies it to the seventh), *Alveolus*, *Choanæ* (in the
modern connotation), *Corpus callosum* (probably first used by
Sylvius, but introduced first to the printed page by Vesalius),
Incus, and *Mitral valve*

We must thus think of Vesalius as trebly equipped for his

task ; firstly, by his own native genius for dissection, developed
with Sylvius at Paris and stimulated by the freedom of

FIG. 65.—Historiated initials from the *Fabrica* of Vesalius. All the
historiated initials in the book show little cupids performing various
anatomical operations.   In many of them Vesalius seems to be
poking fun at his students.
  A.   Vivisection of a pig (see also Fig. 73).   The trachea is
being opened.   Vesalius describes how the animal may be kept
alive by the action of a pair of bellows inserted in the trachea
even if the thorax is opened.   Below to the left a cupid is playing
with a razor.
  B.   Boiling a skull.   On this question of boiling bones see
p. 85.
  C.   Hoisting the body of a dog on to a gallows tree for the
purpose of dissection.
  D.   A "resurrection party".

teaching at Padua ; secondly, by the current attitude toward
the human body exalted by contact with Renaissance Art ;

thirdly, by an admirable education, according to the standards of the time, directed along humanist lines by Günther, one of the ablest medical Humanists of the day For the complete understanding of the *Fabrica* we must remember, too, that Vesalius was a man of superabounding energy who rejoiced in the spoken word. Much of the book is *spoken Latin*, and must be read as though delivered at lecture.

In estimating the anatomical standpoint of Vesalius, too much has sometimes been made of his " anti-Galenism ". In this connexion it must be remembered that in the very years in which he was most busily occupied on the *Fabrica* the leisure of Vesalius was largely devoted to editing Galenic works. His *Tabulæ sex* and his edition of Günther's *Institutiones anatomicæ*, both of which appeared in 1538, certainly represent an early stage in the development of his mind. Not so his editions of works of Galen. In 1541 the great Venice printing firm of Giunta brought out a fine edition of Galen in Latin which is still of value for the study of that author. One of the contributors to it was Vesalius, who edited the works *On the dissection of the nerves, On the dissection of the veins and arteries*, and the great treatise *On anatomical procedure*. Moreover, the numberless references to Galen in the *Fabrica* exhibit the respect of the Reformer of Anatomy for the opinion of the Prince of Physicians. Of a truth it was hardly open to Vesalius to do other than build on Galen. The Galenic works, now accessible in good recent versions to which Vesalius himself had thus contributed, presented by far the best current anatomical accounts. The tone of much of the criticism of Galen is not, it is true, that which a modern writer would adopt. This we must put down to the custom of the time.

§ 3 *On the Supply of Anatomical Material in the Fifteenth and Sixteenth Centuries*

Vesalius has always been regarded as the first modern anatomist to place his study on a firm foundation of observation. It is sometimes forgotten how small his opportunities really were. Thus, between 1537 and 1542 the whole of his experience of the female generative organs was founded on six bodies. Three had to be used for public demonstration. Of the others, one was of a six-year-old girl and had been stolen by a student from the grave. It was in a wretched state of preservation, and could hardly be used except for the study of the bones. There thus remained only two for purposes of private study. Of these one was a pregnant

FIG. 66.—The Uterus and Vagina from the *Fabrica*. The organ is figured as bifid and split longitudinally. The legends may be translated as follows: " AA, BB, sinuses of the *fundus uteri*; CD, a line, somewhat like a suture, projecting slightly into the *fundus uteri*; EE, the thickness of the inner and proper tunic of the *fundus uteri*; FF, a portion of the inner *fundus uteri* projecting downwards from its surface; GG, orifice of the *fundus uteri*; HH, second and external covering of the *fundus uteri* reflected from the peritoneum; IIII, by this we indicate the membranes on both sides which are reflected from the peritoneum and contain the uterus; K, the substance of the *cervix uteri*; L, a part of the neck of the bladder."

woman that had been murdered. Vesalius seems to have been forced to dissect this body very rapidly in the course of a post-mortem examination, conducted for judicial purposes. In accord with this is the inferiority of his description of the pregnant uterus and fœtus. The remaining female corpse was of a woman that had been hanged. It is on her that the anatomy of the female generative organs in the *Fabrica* is substantially based (Fig. 66). Small wonder that Vesalius followed Galen in frequently drawing his conclusions from the bodies of animals !

This is the appropriate place to discuss the method of obtaining anatomical material during the periods of the Middle Ages and the Renaissance. Access had improved since the times of Mondino. At Bologna in the fourteenth century teachers were appointed by the students, and they were bound, according to the Statutes of the University, to dissect such corpses as were brought to them for the purpose. In 1319, during Mondino's lifetime, we hear of students at Bologna being prosecuted for body-snatching. Guy de Chauliac, who was a student at Bologna later in the century, says that his teacher often dissected. The bodies seem usually to have been those of criminals. No formal permission was given save perhaps for post-mortem examination in the course of legal processes. Gradually, however, the authorities came to wink at the proceedings.

Early in the fifteenth century (1405) Padua was incorporated in the Republic of Venice where the rule was less under ecclesiastical control than at Bologna, so that even the skeleton could be adequately studied (cf. p. 85). A Paduan surgeon (Bertapaglia) tells of having witnessed dissection in 1439 and 1440 and Montagnana (p. 95) in 1444 had seen fourteen post-mortems there. The situation throughout Italy was doubtless eased under the Papacy of Sixtus IV (1471–84) and more so under Clement VII (1523–4). Yet later Rome itself became something of an anatomical centre, though less important than Padua or Bologna.

From the early part of the fifteenth century onward the supply of bodies for dissection or post-mortem examination seems to have been fairly steady. To obtain them was doubtless then, as now, largely a matter of tact and judgment. Private workers and practitioners such as Leonardo da Vinci (p. 91), Michelangelo (p. 90), Charles Estienne (p. 100), and Antonio Beniveni (p. 104), seem to have been particularly successful. The last expresses his surprise at being refused a post-mortem, but naturally post-mortem examination was always easier to obtain than permission for actual anatomical dissection.

The professed teachers of Anatomy usually had more difficulty. They were naturally more suspect. But then, as now, the difficulty of obtaining permission for a post-mortem examination was less with the educated than with the ignorant classes. Thus Vesalius tells us that at Louvain he had no difficulty in obtaining permission to open the body of the daughter of a noble family. Post-mortem examination, indeed, is one thing, dissection another. There is no doubt, however, that by the sixteenth century the situation had improved. The embarrassments of Berengar (p. 97) at Papal Bologna were not greater than those of Vesalius at the capital of France or in Venetian Padua. Vesalius had sometimes to rely on body-snatching at both places (Fig. 65). Eustachius at Rome must have used a great number of bodies for his work, which, as a professor at an ecclesiastical college, was performed with the knowledge of the clerical authorities. From about the middle of the sixteenth century the advance of Anatomy in Italy does not seem to have been greatly checked by lack of material.

§ 4   *The Seven Books of the Fabrica of Vesalius*, 1543

We now turn to examine some details of the anatomical masterpiece of Vesalius. We may remind the reader that this book is not only the foundation of modern Medicine as a Science, but the first great positive achievement of Science itself in modern times. As such it ranks with another work that appeared in the same year, the treatise of Nicholas Copernicus *On the Revolutions of the Celestial Spheres*. The work of Copernicus removed the earth from the centre of the Universe : that of Vesalius revealed the real structure of man's body. Between the two they destroyed for ever the favourite mediæval theory of Macrocosm and Microcosm (p. 65).

But the work of Copernicus is one of close and subtle reasoning. It is hardly a great exposition of what we now call the "Experimental Method". That of Vesalius is a vast and not ill-arranged collection of new observations.

It more nearly resembles a modern scientific monograph than does the treatise of Copernicus. Apart from its interest for our particular theme, the work of Vesalius is of high philosophic value as the first great original treatise involving a large amount of observation in any department.

Any attempt to treat the great work of Vesalius comprehensively would demand a volume. We can but glean from it. There are, in fact, two books that appeared from the pen of Vesalius in 1543. In addition to the famous *Fabrica* there was issued also by him an *Epitome* of the work intended for those who were not students of medicine. The figures in the *Epitome* differ slightly from those in the *Fabrica*. We shall treat the two works together basing our remarks on the *Fabrica*.

Vesalius *On the fabric of the human body* is divided into seven books. The first is devoted to bones and joints, the general classification of which is taken direct from Galen. The first bone to be described is the cranium. Those who have not examined the work of Vesalius may be surprised to find a classification of skulls into long, broad, round, etc. These types are figured, and are much the same as those distinguished by modern anthropologists. The sphenoid bone is figured for the first time. A lacuna which appears in it occasionally is described by him, and has since been called the *Foramen of Vesalius*. The Incus and Malleus also appear for the first time in figures, but the Stapes is omitted.

Vesalius contradicts Galen in denying that man has a separate Pre-maxillary bone and he contrasts him in this respect with the dog (Fig. 67). Oddly enough, his figure of the Hyoid bone is probably actually taken from the dog. Vesalius has admirable representations of the vertebræ in the different regions, distinguishing their types, comparing them with those of the ape, and bringing out the salient features in the spinal curves. His figures of the ribs are less satisfactory, and contain a considerable number of errors.

The scapula is not among the best figures of Vesalius. He compares it, too, with that of the dog. The sternum is better. It is described in only three parts, instead of segmented into seven as in the *Tabulæ sex*. He has a good and extensive description of the clavicle, of which the separate cartilages are described. The long bones of the upper arm are too short and rather clumsily rendered but the hand and wrist bones are excellent. Vesalius was the first to give any tolerable description of the carpus. He describes, however, a very minute extra bone at the base of the fifth metacarpal, which, he says, only exists occasionally. He shows, in opposition to Galen, that the bones of the hand contain marrow.

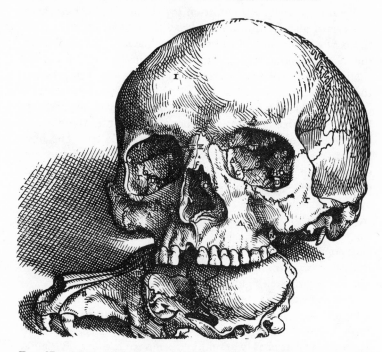

FIG. 67.—Skull of man and dog from the *Fabrica*. The figure is used by Vesalius to illustrate the point, among others, that in the human skull there is no separate premaxillary bone and that in this point it differs from that of the dog.

The figures of the pelvic bones are good and mark a great advance on his *Tabulæ sex* of 1538. The same criticism as was made of his descriptions of the long bones of the upper limb applies also to those of the lower. In the foot, in the angle between the tubercle of the fifth metatarsal and the cuboid bone, he regularly figures a small round bone corresponding with that which he describes in the hand. This bone, in fact, has no existence. Several instances of the separation of the tuberosity of the fifth metatarsal have, however, been recorded, and the separate portion named *Os Vesalianum*. It is a great pity that the only structures named after Vesalius should record his errors in mistaking the abnormal for the normal.

Among the least satisfactory of the descriptions of Vesalius is that of the laryngeal cartilage, perhaps taken from the dog, whilst among the greatest glories of his book are the three exquisite cuts of complete skeletons drawn in dramatic attitudes (Figs. 102–4). He makes these dry bones live.

Before leaving the account of the bones by Vesalius, we may recall the fact that a skeleton prepared by him is still extant. In 1546 Vesalius passed through Basel, where the *Fabrica* had been printed three years before. His reputation was now firmly established. He had with him a skeleton, on which he was invited to demonstrate. On leaving the town he presented it to the University, in whose charge it remains to this day. It is the oldest anatomical preparation in existence.

The second book of the *Fabrica* is devoted to the muscles. The great series of muscle figures (Figs. 105–11) rivals that of the three skeletons as a triumph of anatomical illustration. One of the poses, we note, is traditional, and has been traced to a diagram of Mondeville (compare Fig. 109 and Plate XD). To appreciate the muscle figures of the *Fabrica* they should be examined in conjunction with the seven similar full-page figures in the *Epitome* (Figs. 112–16) and with the works of Berengar of Carpi (Fig. 52), of Estienne (Fig. 57), of Canano (Plate XIV), and of Eustachius (Fig. 74).

It will be seen that the great muscle figures of Vesalius are immeasurably superior to those of Berengar, who, however, had set himself a task somewhat similar to that of Vesalius, and had employed a by no means contemptible artist. Estienne's work is more complete than that of Berengar but utterly inferior to that of Vesalius. On the other hand, the method adopted by Canano has certain advantages and undeniably brings out points not well illustrated by Vesalius. When we compare Eustachius with Vesalius, we are comparing men of similar intellectual rank, though of very different anatomical outlook. Eustachius is picturing dead Anatomy, Vesalius living. From the point of view of distinguishing anatomical details, however, Eustachius is often, perhaps usually, the superior, showing for instance the nerve supply, a matter omitted by Vesalius.

The figures of Vesalius exhibit the muscles in a state of contraction, and almost invariably suggest movement and activity. In one very interesting figure there is an extension of the *rectus abdominis* upwards over the upper ribs as in monkeys (Fig. 107) ; in the text he refers to this difference between the human and simian anatomy. Vesalius made constant attempts to determine the actual mode of action of each muscle and tendon, and he has an entertaining figure illustrating the working of the annular ligaments of the foot. He describes a seventh muscle of the eye, the *choanoides*, as a normal human structure. It is, in fact, only found in animals, as was later shown by his successor, Casserio (p. 161). On the other hand, Vesalius failed to distinguish the human *levator palpebræ*.

The third book is devoted to the vascular system (Fig. 68). The several diagrams of the veins which open this book contain some errors, and the book is perhaps on the whole the least satisfactory of all. The treatment of the vascular system is, however, immeasurably superior to that of Estienne, who attempts a like scheme. The figure of the azygos vein by Vesalius is inferior to that by Eustachius and the general diagram of the arterial system provided by Vesalius is poor. He gives us, however, a striking diagram of the veins of the brain. Of the pulmonary vessels, both arterial and venous, there are remarkable representations exhibiting their complete course with heart and lungs dissected away. It is difficult to see how such preparations could have been made save by a method of injection. The book ends with a large figure of the vascular system and a page of diagrams of various organs (Fig. 68). The latter, says Vesalius, are to be cut out and pasted on to the former. This is the first instance in which the type of demonstration by moveable layers is adopted. It

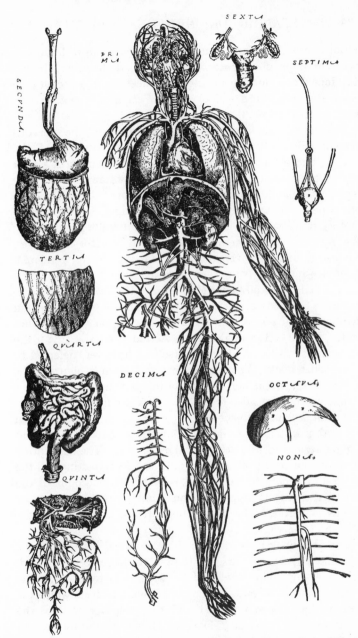

FIG. 68.—A page from the *Epitome*. The figures are most of them reproduced in the *Fabrica* from the same block.

is not a useful method, and has since fallen into desuetude
save for popular purposes. In our period, however, it was
developed occasionally in fugitive sheets and especially by
Hans Remmelin (1583–1630) in his *Catoptron microcosmicum*
first printed in 1613.

The fourth book of the *Fabrica* treats of the nervous
system (Fig. 69). The figures rank below those of the
osseous and muscular systems. The book opens with a poor
view of the base of the brain. The olfactory tract is defectively
represented and the olfactory bulb cannot be seen ; the
optic nerve is wrongly drawn ; the origins of the oculomotor, trochlear and abducent are all falsely represented ;
the roots of the trigeminal facial and auditory are very
confused, and the remaining roots are badly rendered.
The classification of nerves adopted is into seven pairs,
as in Galen (p. 56) and Mondino. The *pons* is unrepresented. The general surface of the cerebrum and cerebellum
are fairly well portrayed. There follows a very crude figure
of the brain and cranial nerves viewed from the side ; the
general discussion of the brain itself is, however, deferred to
the seventh book. We notice a clear and excellent figure of the
recurrent laryngeal nerves, exhibiting the difference in their
course on the two sides (Fig. 70). They are figured, however,
from an animal subject. The course had already been
admirably described by Galen (p. 56). In the text Vesalius gives
a good account of the action of these nerves. The description
of the spinal cord itself is poor, and he fails to distinguish the
two sets of roots to the spinal nerves. The account of the
brachial plexus is imperfect, but that of the lumbo-sacral
plexus is better. The sympathetic trunk is described as a
branch of the vagus.

The fifth book is devoted to the abdominal viscera. These
are passed over more cursorily than might have been expected,
but in many points the brief description is excellent. The
book opens with a figure of the abdominal wall from which the
muscles have been removed and the posterior wall of the
Rectus sheath exposed. The Great Omentum and the general
lay-out of the intestines are then described and figured. The
intestines and their attachments are excellently shown in a

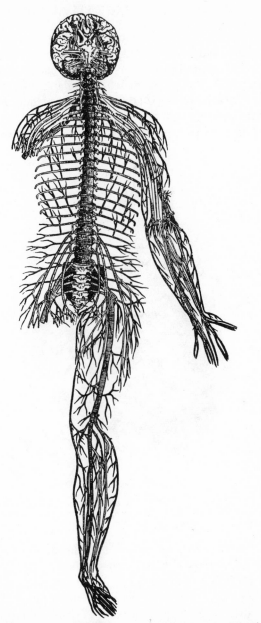

FIG. 69.—The nervous system from the *Fabrica*. An identical figure printed from the same block is to be found in the *Epitome*.

series of figures. Curiously enough, the Vermiform Appendix
is not mentioned in the text, though it is clearly portrayed
at least three times in the figures (Fig. 68 *quarta*). There is a
good figure of the great mesenteric gland, the ' Pancreas
Aselli '. Great emphasis is laid on the gall bladder in the true
line of mediæval tradition. The treatment of the stomach, liver,
spleen, and kidneys is inadequate. The description of the male
generative organs, their form, blood supply, and relations, is

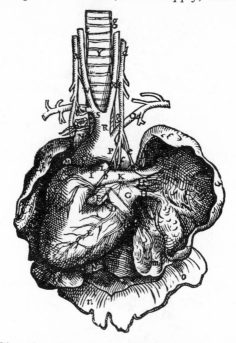

Fig. 70.—Dissection of heart and adjacent parts showing recurrent
   laryngeal nerves from *Fabrica*. Compare the description of
   Galen as represented in Fig. 29.

fair (Fig. 63). The description of the female system is, however, mediæval and full of errors (Fig. 64). These errors
are set forth also in the figures of Vesalius, and through
them they have been perpetuated in popular Anatomy.
Figures of the female generative organs almost exactly
similar to those of Vesalius are, in fact, circulating to
this day in popular works. The uterus is represented as

slightly bifid (Fig. 66) and is compared with that of a cow and of a bitch. The account of the embryo is negligible. We have already discussed the reason for the very imperfect account of the female generative organs (p. 120). There is a crude discussion of the kidney, in which the pelvis is described and represented as divided in two by a sieve-like structure.

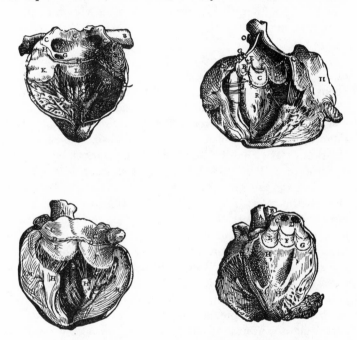

Fig. 71.—Dissections of the heart from the *Fabrica*. The various valves are shown and also the interventricular septum. Vesalius remarks on the pits in the septum and says they are imperforate and a bristle cannot be passed through them. The point is of importance in connexion with the interpretation of the Galenic physiological system. See p. 132.

The idea is taken from Galen and exhibits no advance on Mondino (p. 81). In discussing this organ the sieve is represented diagrammatically.

The sixth book contains a description of the heart and lungs. It opens with a figure which appears in many subsequent anatomical works. In it the chest is opened from the

side, exhibiting the Phrenic Nerve. The lungs are very briefly
treated, and in the figures the right lung is divided into only
two lobes. The account of the heart is good and interesting.
The physiology of Galen (p. 58) is generally accepted as regards
the heart's action, although the position is somewhat altered
in the second edition of 1555. Interest naturally concentrates
on the description of the septum. Vesalius says that he has
tried to put bristles through the pits there (Fig. 71), but
has failed.

The seventh book treats of the brain and includes a series
of very fine figures of an absolutely pioneer character. It
opens with an excellent representation of a head from which
the calvaria has been removed. The middle meningeal artery
can be seen meandering across the surface of the dura. This is
followed by a series of dissections and horizontal sections of
the brain. They exhibit admirably and clearly a whole series
of structures (Fig. 72). In the ventricles can be seen quite
distinctly the *caudate nucleus*, the *thalamus*, the *stria terminalis*,
the *choroid plexus*, the *fornix*, and the beginning of the
hippocampus major. In the substance of the brain we may
distinguish a general division between white and grey matter,
the *internal capsule*, the *caudate nucleus*, and the *lenticular
nucleus*, showing the division into *putamen* and the *globus
pallidus*. There is also an excellent account and view of the
mid brain involving the pineal gland, *pulvinar*, the *corpora
quadrigemina*, and the *superior* and *middle cerebral peduncles*
extending downwards to the *bulb*. The *fourth ventricle* is well
shown. The cerebellum is lightly treated.

Emphasis is laid by Vesalius on the pituitary gland and its
fossa. Especially he deals with the supposed relation to the
rete mirabile, the existence of which he does not wholly reject.
His figure of this structure, he says, is made to fit the description
of Galen ! This shows how deeply steeped he was in the Galenic
physiology. There follows an extremely bad account of the
internal structure of the eye (Plate XIIB). Vesalius had no
idea of the functions of the different parts of this organ.
Following all mediæval anatomists, he regards the *crystalline
lens* as spherical and he places it in the centre of the globe
of the eye (cf. Fig. 49). He considers that it performs the

functions which we now ascribe to the retina. These errors were not corrected until 1583, when the Swiss anatomist Felix Plater (1536–1614), in a work published, like the *Fabrica*, at Basel, but exactly forty years later, began to put the physiology of the eye on a modern basis. Minor corrections in the anatomy of the eye had already been made

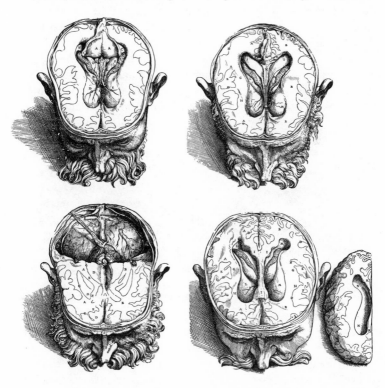

FIG. 72.—A series of dissections of the brain, from the *Fabrica*.

at that time by Columbus. The description of the other organs of special sense by Vesalius is wholly inadequate.

The work terminates with a very interesting little chapter *On the dissection of living animals*, which is worth translation *in extenso*. It deals with the methods of physiological experiment available at the time. Comparatively little

advance is exhibited on the methods of Galen, but the subject
is skilfully and succinctly handled.

Among the experiments that Vesalius enumerates are
excision of the spleen—the loss of which he showed was
consistent with life—and cessation of voice with the cutting
of the recurrent laryngeal nerves. He demonstrated that
longitudinal section of a muscle interfered little with its
function, but cross section produced disability in proportion
to the injury. Such experiments had been performed by
Galen in antiquity who had also reached the same conclusions
as Vesalius that it is through the spinal cord that the
brain acts on the various muscles of the limbs and trunk.
Vesalius repeats Galen's experiments on section of the Spinal
Cord (p. 60). More original is his observation that nervous

FIG. 73.—A pig prepared for operation, from the *Fabrica*.

impulses pass not through the sheath but through the
substance of the nerve. Even more striking are his
experiments on respiration. Here he showed that even though
the thoracic wall be pierced the animal may be kept alive
by aerating the lungs by means of a bellows and that a
flagging heart may be revived by similar means.

The work of Vesalius was often reprinted. After the first
publication, the *Fabrica* appeared in at least twenty-five
editions between the years 1543 and 1782, being issued from
the presses at Augsburg, Basel, Cologne, Ingolstadt, Leyden,
London, Nuremberg, Paris, and Venice. The figures of
Vesalius were copied and plagiarized from the beginning.
Among the more shameless were Ambrose Paré (1510–90),

who reproduced them without acknowledgment in 1551 and afterwards, and Helkiah Crooke (1576–1635), who plagiarized them in 1615. The latter, particularly, adds insult to injury by accusing Vesalius of having slighted Galen !

§ 5   *Eustachius, Rival of Vesalius, flourished* 1550–74

A name that has not been sufficiently recognized in the history of Anatomy is that of Bartolomeo Eustachio (1520–74). Eustachius belonged to a different school to Vesalius, and does not seem to have been connected with the North Italian Universities. He practised in clerical circles in Rome, and was a great upholder of Galen. Eustachius resembled Leonardo in that his anatomical achievement was very much greater than the influence which he exerted, and this for a similar reason. His work—with certain insignificant exceptions—was not published during his lifetime, and nearly all the text is lost. The splendid copper plates that he had prepared were, however, discovered in the early eighteenth century, and were presented by Pope Clement XI to his physician Lancisi (1655–1720), who published them in 1714 with his own explanations. For purposes of study, the edition issued at Leyden in 1744 with the legends of B. S. Albinus (1697–1770) is perhaps more valuable. Had these plates of Eustachius appeared in 1552, when completed, his name would have stood by the side of Vesalius as one of the founders of modern Anatomy. The plates of Eustachius are less beautiful than those of Vesalius. They present dead and not living Anatomy. They are, however, more accurate, and they contain such a multitude of discoveries that for originality Eustachius has only Leonardo and Vesalius as superiors among modern anatomists. We may rapidly run through these plates, discussing some of the more salient points.

The figures of the kidney by Eustachius are among the few published during his lifetime, and are superior to those of Vesalius. In discussing them he attacks Vesalius for having represented the kidney of a dog in place of that of a man. His treatment of the kidney introduced the study of anatomical

136

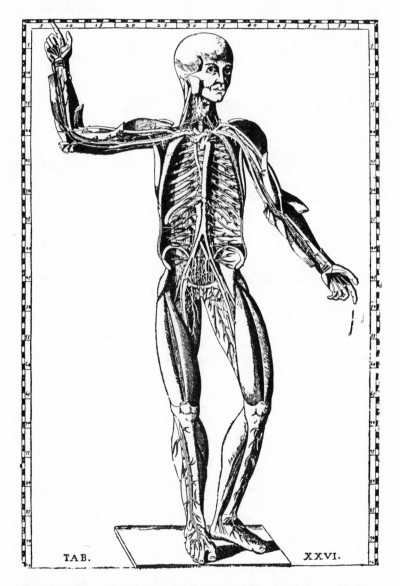

TAB.   XXVI.

FIG. 74.—From Bartolomeo Eustachio, *Tabulæ anatomicæ*, edited by
J. M. Lancisi, Rome, 1714. Plate showing vascular system
with general relations of blood vessels to muscles.

variations. The subject was hardly considered till modern times, but Eustachius applies it to many other parts, kidney, azygos vein and veins of the arm (Fig. 75), brachial artery, innominate artery, gastric blood supply, etc. He has excellent

FIG. 75.—From Bartolomeo Eustachio, *Opuscula Anatomica*, Venice, 1563. I and II, system of veins in the arm of man ; III and IV, veins in the arm of an ape ; V, veins in the arm of a dog ; VI, right auricle and ventricle opened, showing musculi papillares, fossa ovalis, annulus ovalis, Eustachian valve, and coronary valve.

figures of the ear ossicles, and the *tensor tympani* in man and in the dog. Oddly enough, however, he has no figure of the tube to which his name is now attached. The description of that structure is to be found in his work *Examination of the organ of hearing*, which was published during his lifetime in 1562. The Eustachian tube was known to Alcmæon as long ago as 500 B.C., and Aristotle also refers to it. There is, moreover, evidence that two contemporaries of Eustachius, Vesalius and Ingrassias (1510–80), were acquainted with it. Eustachius investigated also the internal ear, and we owe to him the term *Modiolus*.

During his lifetime Eustachius published a figure of the heart showing the *fossa ovalis* (Fig. 75), which he claimed as a discovery. It had, however, been described by Sylvius and is shown in a figure of Vesalius (Fig. 71). He rendered the cardiac vessels well. Eustachius displays quite correctly the relations of vein, artery, and bronchus in the lung in a manner which was not even attempted by Vesalius. The drawings by Eustachius of the abdominal viscera are about equal in accuracy to those of Vesalius, though inferior in beauty. Eustachius, however, shows a large number of abdominal lymph glands, which Vesalius had not observed. The figures by Eustachius of the female generative organs escape some errors into which Vesalius had fallen. The figures of the nervous system by Eustachius are, in general, inferior to those of Vesalius. Nevertheless, the glory of the whole Eustachian collection is a truly magnificent drawing of the Sympathetic System (Fig. 77). We doubt if any better and clearer portrayal of the connexions of that system as a whole had been set forth until our own day. It is a really great anatomical figure, and is by itself sufficient to place Eustachius in the front rank of anatomists. The same remarkable figure shows the base of the brain, with the roots of the cranial nerves far more clearly and accurately rendered than by Vesalius. The *pons*, too, is shown better than by Varolius, whose name is now attached to it (Fig. 78).

The muscle figures of Eustachius are very stiff and ugly compared with those of Vesalius, but are much superior in their detailed accuracy and in the fact that the nerve and blood supplies are shown (Fig. 74). The general diagram of the arterial and venous systems are at least equal to those of Vesalius. Eustachius has also a series of muscle figures without the nerves ; among these and beyond anything that we have in Vesalius are the representations of the muscles of the face and of the laryngeal apparatus. In his treatment of the organ of voice, indeed, Eustachius easily surpasses his great rival (Fig. 76). The work

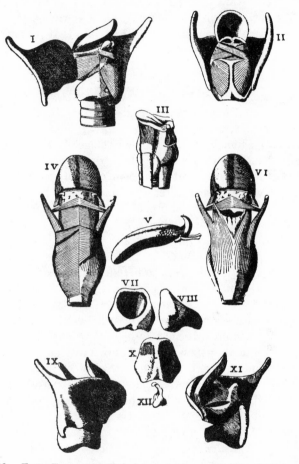

Fig. 76.—From Eustachio, *Tabulæ anatomicæ*, edited by J. M. Lancisi, Rome, 1714, showing dissections of larynx.

terminates with a series of figures of bones inferior to the Vesalian plates at almost all points. It is interesting to see here a figure of the skull of an ape placed by the side of the human structure. It has not been generally observed that

Eustachius described the thoracic duct nearly a century before Jean Pecquet (1622–74) wrote his *New Anatomical Observations* (1651), with his account of the *receptaculum chyli*. Eustachius gives no figure of the thoracic duct, but he says that in horses it is like a white vein, that it brings chyle to the heart, that it has on it a half-moon shaped swelling, and that it opens into the internal jugular veins.

§ 6   *The Followers of Vesalius*, 1550–90

The tragic story of Michael Servetus (1511–53, Plate XV) hardly affected the course of Anatomy, save in so far as his doctrine of the lesser circulation may have influenced Columbus and through him Harvey. It has not usually been observed that Servetus practically reverts from the physiology of Galen to that of Erasistratus, adopting two kinds of spirit instead of three. It will be remembered how near Erasistratus came to discovering the circulation (pp. 32–3). Servetus, like Vesalius, was a favourite pupil of Günther. It is with the school of Vesalius that we now have to deal.

In the first line comes Realdus Columbus (1516 ?–59), of whose attainments very different estimates have been formed. Columbus was assistant to Vesalius, and is mentioned by him with generosity. He taught Anatomy at Padua in succession to Vesalius, and his anatomical work appeared posthumously in 1559. It is essentially a " textbook ", containing few original observations, but better arranged and easier to read than that of his great teacher. It loses greatly, however, by being devoid of illustration.

The textbook of Columbus, though retrograding at some points from the Vesalian standpoint, yet often exhibits real advances. We may leave aside the question to what extent these discoveries were made by Columbus himself. He certainly deserves credit for having displaced the lens from its age-old position in the centre of the globe of the eye, where even Leonardo and Vesalius had left it. Columbus is particularly strong on regional Anatomy, and his descriptions of the Mediastinum, of the Pleura, and of the Peritoneum are far ahead of anything that had preceded him. The modern use of the term *Pelvis* dates from Columbus, as does also *Bregma*, a word

PLATE XV

SERVETUS IN PRISON
From the statue by Clothilde Roche at Annemasse,
Haute-Savoie, France (four miles from Geneva).

[face p. 140

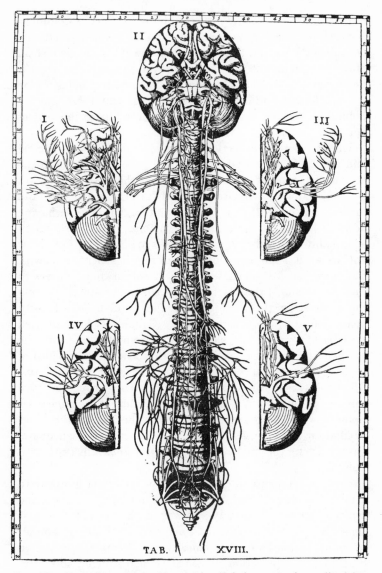

FIG. 77.—From Bartolomeo Eustachio, *Tabulæ anatomicæ*, edited by
J. M. Lancisi, Rome, 1714.   Plate showing the base of the brain
and the sympathetic nervous system.   The figures of the base of
the brain are better than those of Varolio (see Fig. 77), and the
figure of the sympathetic is one of the best and clearest of that
system and its connexions that has ever been produced.

he drew from Aristotle. In the chapter on the larynx Columbus accused his master Vesalius of having demonstrated and described the larynx, the tongue, and the eye of the ox instead of the human organs. In the case of the larynx, however, it is much more likely that Vesalius used the dog, if not working on the human subject. The charge of fraud brought by Columbus against the man to whom he owed so much does not read pleasantly.

The chapter on vivisection in the work of Columbus is good and clear. In it we naturally turn to the paragraph on the action of the heart, and we there find the observation that cardiac systole is synchronous with arterial expansion, and cardiac diastole with arterial contraction, the reverse having been the belief of more ancient writers. This observation is probably really his own, and greatly to his credit. Columbus notes also that the pulsation of the brain is synchronous with the pulsation of the arteries. The attention that he gives to the movements of the heart and lungs is important as showing the interest in these subjects at the Paduan school forty years before Harvey came there. Columbus demonstrated *experimentally* that the blood passes from the lung into the pulmonary vein. It is well known that as regards the idea of a lesser circulation Columbus was preceded by Servetus. The book of Servetus appeared in 1553. The entire issue, save three copies, was burned, along with its author, in the same year. Columbus has been accused of taking the idea of the lesser circulation from Servetus, but the evidence is inadequate. In any event the credit of the *demonstration* still rests with Columbus.

It is important to note that Columbus does not hestitate to attack Aristotle, though Padua was, and long remained, the centre of Aristotelian study and was, in this respect, the most conservative of the Italian Universities. Harvey himself bears many marks of that conservative Paduan Aristotelian tradition, for opposition to which Galileo suffered so sorely.

Gabriel Fallopius (1523–62) was another pupil of Vesalius at Padua, and succeeded Columbus as a teacher there. His *Anatomical Observations*, printed in 1561, contain descriptions of the tubes named after him, and of the ovaries and of the

round ligaments.   Fallopius gave the scientific names that
they now bear to the *Vagina* and *Placenta*.   He introduced into
anatomy the terms *Cochlea, Labyrinth, Hard and Soft Palate*,
and *Velum Palati*.   He rendered the first account of the
Chorda Tympani, of the semi-circular canals, and the sphenoidal
sinuses and of the ' aqueduct ' named ' Fallopian '.   His
descriptions of the trigeminal, auditory, and glosso-pharyngeal
nerves were the best up to their time.   He described the
trochlear nerve as a separate root.   His work called forth a
famous rejoinder from Vesalius.   Fallopius was a very effective
teacher, and by his character and manner impressed himself

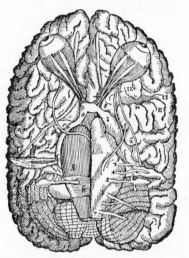

Fig. 78.—Base of brain from Costanzo Varolio, *De nervis opticis
nonnullisque aliis præter communem opinionem in humano
capite observatis*, Padua, 1573.   In the accompanying description
we read that the part bearing the letter *h* is a *Processus transversalis
cerebri, qui dicitur Pons*.

more on his contemporaries than he has on subsequent
anatomical literature.   His early death limited his output,
but we shall trace his influence in the work of his pupils,
Coiter (p. 148) and Fabricius (p. 153).

Inferior to these teachers of the direct Vesalian line was a
professor at the rival school of Bologna, one Constanzo
Varolio (1543–78).   His only important anatomical work
appeared without his permission in 1573 and treated of

the base of the brain. It contains a few crude figures displaying the *pons* (Fig. 78), which is still called after him, but is much worse rendered than in the figure of Eustachius. Another Bologna professor, Giulio Aranzi or Arantius (1530–89) produced a book *On the human fœtus* in 1564 and *Anatomical observations* in 1587, in both of which works anatomical discoveries of some significance were announced. He gave the first adequate printed account of the gravid uterus, and finally dispelled the idea of a human cotyledonous placenta. Many had believed in this before his time, and, as we have seen, a cotyledonous human placenta figures in the magnificent drawings of Leonardo. Arantius gives a noteworthy description of the Anatomy of the fœtus, by far the best up to his time. He especially examined the fœtal heart, and saw the *ductus arteriosus* and the *foramen ovale*. He paid great attention also to the vascular system of the adult and gave a good description of the lesser circulation which did not, however, advance beyond the standpoint of Columbus. He described the little nodules of cartilage in the semi-lunar valves to which his name is now attached. Vidus Vidius (died 1569) had already described them in a work which however, was not printed till 1611. This book of Vidius contains the first published figure of the Sympathetic. It is, however, far inferior to that of Eustachius, which was prepared much earlier but published much later. Vidius is commemorated in the Vidian nerve, artery, canal, and vein.

Special attention to the heart of both fœtus and adult was paid about this time by the Roman anatomist, Archangelo Piccolomini (1526–1605), who gives very poor figures of the *ductus arteriosus* in his *Anatomical Lectures* published at Rome in 1586. This work is of little worth, but has acquired some interest from the fact that Harvey largely depended on it. Another inferior contemporary anatomist whose textbook Harvey studied was the Provencal, André du Laurens, (died 1609) of Montpellier. His textbook, which appeared in 1595, was the most popular of its time, and was frequently reprinted. It was well illustrated. Du Laurens took most of his figures from Vesalius, and made few observations on his own account. Among the few were those on the skeleton of the child at different ages. He gives a figure of the *cauda equina* drawn just like a horse's tail. The term *cauda equina*, which we owe to him, is said to be the translation of a Hebrew term found in the *Talmud*. He is also responsible for the terms *Optic Chiasma* and *Phalanx* in their modern application.

§ 7 *The Early Comparative Anatomists, 1540–1600*

The earlier anatomists, as we have seen, relied largely on the dissection of animals, from necessity rather than from choice. Vesalius began as a boy to dissect small animals. As a man he was the first to draw systematic parallels between the structures of animals and men. He compared particularly simian with human anatomy, and refers specifically to various structures in the ape, among them the lumbar vertebræ and their processes, and the sacrum and coccyx ; the muscles of the thorax, arm, thigh, hand, and foot ; the spinal nerves ; the lung ; the omentum, mesentery, and colon. He compares the structures of the tailed and tailless monkeys, and he invokes a large number of other animals, e.g. oxen, goats, sheep, dogs, and cats. The main object of the comparative studies of Vesalius was to show that the anatomical writings of Galen described the structure of animals not of man. He proved his point up to the hilt.

Vesalius had been accustomed to see animals dissected in the theatre of his teacher, Sylvius, at Paris. Perhaps in response to the charge of Vesalius that he was not accustomed to dissect the human body, Sylvius himself prepared a small work, *Observations in the dissection of various bodies*. It was issued just after the death of Sylvius in 1555. It opens with the description of four human bodies and continues with an account of the dissection of the bodies of a number of animals —the monkey, sheep, cow, pig, dog, horse, stag, lion, and some mammal, which he calls *trocta*, that I am unable to identify. The anatomical descriptions in this work of Sylvius are very imperfect and superficial.

La peincture de l'Embryon d'vn Marſouin.

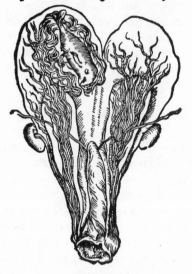

Fig. 79.

La peincture de l'Oudre, que les Latins nomment Orca ou Orcynum.

Fig. 80.

Two figures from Pierre Belon, *Histoire naturelle des estranges poissons marins*, Paris, 1551.

Fig. 79.—The uterus of a porpoise opened to show fœtus attached by umbilical cord to placenta.

Fig. 80.—Grampus and newly born young. The fœtus is still surrounded by its membrane and the afterbirth is in process of extension.

The immediate successors of Vesalius—Fallopius and Columbus—both dissected animals, but showed little inclination to adopt a definitely comparative standpoint. Later in the century, however, a number of workers devoted themselves to comparative studies, and it was they who determined the general progressive character of the Anatomy of the time. It will be impossible to consider all of these writers. Among the most typical were Coiter, Fabricius ab Aquapendente, and Casserius. At the same time Belon and Rondelet, the French naturalists, were engaged in the systematic study of marine forms.

Pierre Belon of Le Mans (1517–64) produced in 1551 a valuable little treatise on Fishes. A book by him of the same year on Birds shows the skeleton of a man and of a bird

Fig. 81.—*Mustelus (Galeus) lævis*, and young from Guillaume Rondelet, *De piscibus marinis*, Lyons, 1554.

placed side by side to exhibit their homologies. It is a most striking figure and the earliest of the kind in a printed book. Belon was the first to examine the placenta of the porpoise and to figure the creature in its mother's womb bearing characteristics of a mammalian fœtus (Figs. 79, 80). Guillaume Rondelet of Montpellier (1509–66) was a pupil of Günther, and later became professor of Anatomy at Montpellier. He is the " Rondibilis " of Rabelais, who was himself one of the medical humanists. This Rondelet is said to have been so enthusiastic an anatomist that he dissected the body of his own son, but he made no contribution to human Anatomy. Rondelet produced, however, in 1555 an admirable and beautifully illustrated work *On marine fishes*, in which he described and figured all the marine creatures of the Mediterranean that were then known. Among them is the placental dogfish of Aristotle, with the young attached to the mother's body by a navel string (Fig. 81). Belon and Rondelet based their work on the Aristotelian biological texts. Their books are the two earliest on scientific Zoology.

More a comparative anatomist, as distinct from a zoologist, was the Hollander, Volcher Coiter (1534–76 ?). He studied under Fallopius at Padua, under Arantius at Bologna, under Eustachius at Rome, and under Rondelet at Montpellier. With the traditions thus acquired of accurate structural investigation on the one hand, and zoological interest on the other, he naturally turned to Comparative Anatomy, a subject of which he was the first systematic exponent. Settling in Nuremberg, he brought to Germany the scientific methods of his teachers. In the years 1573 and 1575 he published volumes containing a great number of original and important observations. They are the first books definitely devoted to comparative studies, and they place him very high among the great anatomical pioneers.

The works of Coiter, which are rare, are as concise as they are original, and are admirably illustrated by his own hand. He advises anatomists to read nothing on Anatomy save the great works of Galen, the *Fabrica* of Vesalius, the *Anatomical Observations* of Fallopius, the *Examination of the Anatomical Observations of Fallopius* by Vesalius, and the writings of Eustachius. Looking back on the literature available in his time, no better advice could have been proffered. He constantly urges the comparison of human Anatomy with that of beasts as an occupation worthy of a philosopher.

Coiter gives a remarkable account of the development of the hen's egg, and the formation of its various parts. With the exception of a few observations by Albertus Magnus (1206–80) in the thirteenth century, this is the only work of its kind since Aristotle. Coiter opened incubated eggs day by day, and the descriptions of his findings must have acted as a guide to the subsequent researches of Fabricius. So far as modern times are concerned, Coiter is unquestionably the father of Embryology. He is the first to give figures of the skeleton of the fœtus. He shows and gives admirable descriptions of the skeleton of a miscarriage of six months and of a much earlier abortion, and in each case he notes the state of ossification.

More purely in the department of Comparative Anatomy, Coiter gives an excellent drawing of the skeleton of a tailed monkey, and compares it in detail with that of the tailless monkey and with the human subject. He has an excellent description of the organ of hearing, including the tympanum, the ossicles, the *tensor tympani*, the Eustachian tube, the *chorda tympani*, the *aqueductus Fallopii*, the two *fenestræ*, the labyrinth, the *cochlea*, and the auditory nerve. This is by far the best description of its kind up to the time of Casserius (p. 161). Coiter was the first to publish descriptions of the frontal sinuses and their opening into the nasal cavity, though Leonardo had known and drawn them (Fig. 43). Coiter made observations on the origins of the cranial nerves in correction of those of Vesalius and of Eustachius. He observed the double roots of the spinal nerves and noticed the distinction between white and grey matter in the spinal cord. He realized that the *rete mirabile* is indistinguishable in man, though well developed in the ox. He examined embryonic pigs of eight or ten days and compared them with chicks.

Most interesting are the observations of Coiter on the living heart. The rarity of his works has prevented it from being generally noted that he examined the living hearts of cats, lizards, serpents, frogs, eels and other fish much in the manner of Harvey. He gives a detailed description of his observations on the heart of a new-born kitten, observing that the contraction of the auricles is followed by, and is not contemporary with, that of the ventricles. He made the important observation that the heart is lengthened in systole and shortened in diastole. He noticed that in excised hearts the different parts continue to beat, and that the part in which the pulsation disappears last is the base. He also perceived that pulsation would continue in a small separated part of heart muscle. He observed the difference in the character of the lungs and the different mechanism of breathing between lizards and frogs on the one hand and mammals on the other, and he noticed the air sacs in birds. Coiter examined the poison apparatus in the viper. He gives excellent little

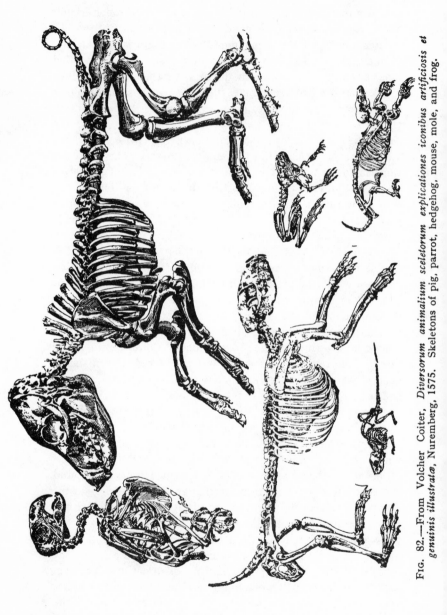

Fig. 82.—From Volcher Coiter, *Diversorum animalium sceletorum explicationes iconibus artificiosis et genuinis illustratæ*, Nuremberg, 1575. Skeletons of pig, parrot, hedgehog, mouse, mole, and frog.

sketches of the anatomies of the tortoise, the hedgehog, and bat, and a fine chapter on the anatomy of birds. He made an attempt to classify mammals on an anatomical basis.

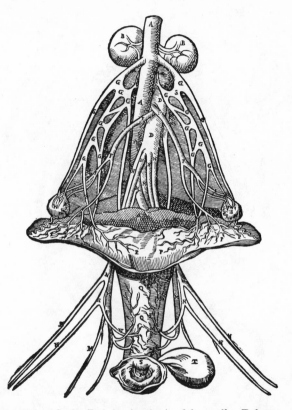

FIG. 83.—From Carlo Ruini, *Anatomia del cavallo*, Bologna, 1598. The uterus and adnexa: AA, vena cava; BB, kidneys; DD, aorta; FF, ovarian arteries; GG, ovarian veins; HH, ovaries (*testiculi*); MM, veins from the hind limbs and uterus; NN, branches of aorta to hind limbs and uterus; PP, cornua of uterus; QQ, body of uterus together with vagina; R, vulva; S, labia; T, bladder.

Coiter's main achievement, however, is a systematic account of the skeletons of a large variety of animals. These are well and accurately figured, and their homologies and affinities carefully described This part of his work is particularly deserving of wider recognition. Coiter gives descriptions of the skeleton of the goat, horse, pig, badger, squirrel, hedgehog, mole, bat, lizard, frog, and tortoise. He has a good chapter on

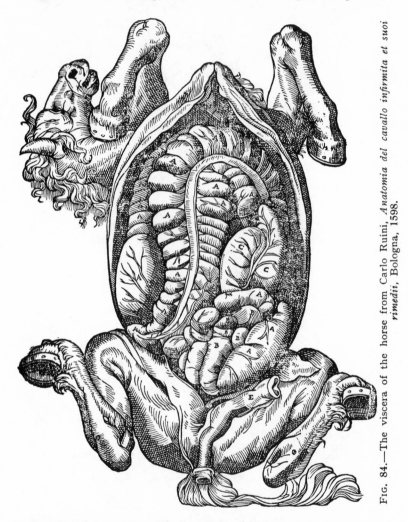

FIG. 84.—The viscera of the horse from Carlo Ruini, *Anatomia del cavallo infirmita et suoi rimedii*, Bologna, 1598.

the skeletons of birds and he reproduces figures of the parrot, cock, woodpecker, crane, cormorant, and other birds. The great rarity of the works of Coiter has constantly militated against their study. Few of his anatomical terms have, therefore, gained currency. One of them is the *Corrugator supercilii*. Even rarer than the works of Coiter is a small treatise by a Neapolitan Franciscan named Germano along somewhat the same lines. Germano's work appeared in Naples in 1625.

Occupying an isolated position is the splendid monograph on the *Anatomy of the horse* by Carlo Ruini of Bologna, published posthumously in 1599. It is the product not of a physician, nor of a professional veterinary surgeon, but of a lawyer. Nevertheless, it does for equine Anatomy a similar service to that which the *Fabrica* of Vesalius had done for human Anatomy ; its truly magnificent figures need not fear comparison with those of Vesalius and of Eustachius, by the side of which they may be placed (Figs. 83–5). The text is no less admirable than the figures ; the description of the eye, ear, intestines, kidneys, and bladder being specially good. Ruini gives a clear account of the structure of the heart and of the mechanism of the pulmonary circulation. His book is the first devoted to the anatomy of an animal, and is one of the finest achievements of the heroic age of Anatomy. The achievement of Ruini is more remarkable, in that he had no forerunner worth mentioning. He keeps strictly to his subject, however, and is not turned aside by any considerations of human or comparative Anatomy. His work is thus purely descriptive, but includes elementary physiological considerations.

§ 8 *Fabricius ab Aquapendente*, 1590–1610

A pupil of Fallopius who exhibited a taste for comparative studies as keen as that of Coiter was the more famous Hieronymus Fabricius ab Aquependente (1537–1619). Fabricius was unquestionably one of the greatest of all teachers of Anatomy. He succeeded his master, Fallopius, and built, at his own expense, the anatomical theatre at Padua, which is still standing. Among his many claims to notice his greatest is perhaps that he taught Harvey. In the year 1604 he was succeeded at his own request by his pupil, Giulio Casserio, and many of his writings appeared after that date. Once relieved of the duties of his chair, he produced in a rapid succession a number of anatomical, embryological, and physiological works of the first rank. These memoirs of Fabricius are characterized by their wealth of large clear illustrations which long remained unexcelled in their particular department. They are copper-plates, not woodcuts. They cover a wide field of embryological and comparative anatomical study. The only figures included by Harvey in his great book *On the motion of the heart* were taken from one of these works of his master. Fabricius ab Aquapendente must not be confused with his German contemporary, Fabricius Hildanus (1560–1634), who made important contributions to Surgery, but had little influence on Anatomy.

The work of Fabricius ab Aquapendente *On the development of the eggs of birds* is a unique document of the highest value for the history of Embryology. Harvey, in his treatise *On generation* (1651) leaned very heavily upon it. Fabricius carried the subject far beyond where Coiter had left it, and elevated Embryology at one bound into an independent science, the importance and interest of which has never since been lost from sight. The work has the great merit of being well and copiously illustrated (Fig. 87). Fabricius here, as elsewhere, exhibits that reverence for Aristotle that we find in Harvey.

Fabricius does not seem to have used a magnifying glass for his work, so that his descriptions of the earlier stages in the development of the chick cannot be expected to excel those of Coiter. For improved observation of these earlier stages, the world had to wait for the work of Marcello Malpighi (1628–94). From the sixth day onward, however, the description and the figures of the chick by Fabricius are, on the whole, excellent.

No less remarkable is the treatise of Fabricius *On the formed fœtus*. This is a magnificent comparative study of the embryo in the more advanced state, and is the first work of its kind. It describes developmental stages in a long series of animals, man, rabbit, guinea-pig, mouse, dog, cat, sheep, pig, horse, ox, goat, and deer among mammals, the smooth dogfish among fishes, and the viper among other creatures. In the figures a good deal of attention is given to the heart, and the *ductus arteriosus* and *foramen ovale* are frequently

FIG. 85.—From Carlo Ruini, *Anatomia del cavallo*, Bologna, 1598.
The surface muscles.

shown (Figs. 88–91). In the text special attention is drawn to
the structural changes in the vascular system incident on birth.
The work contains the best figures up to its time of the human
gravid uterus and membranes and of the human placenta. It
includes a series of fine demonstrations of the course and
relations of the umbilical vessels, and dissections of various
parts of the human fœtus. Even more detailed is the
investigation of the uterus, placenta, membranes, vessels, and
fœtus of the sheep. The book also contains the earliest figure
of the heart of a fish.

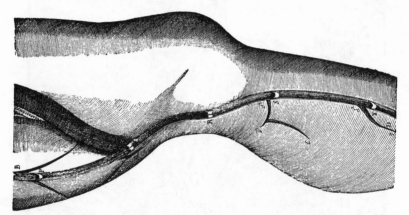

FIG. 86.—Dissection of veins on thigh and leg from Fabricius ab
Aquapendente, *De venarum ostiolis*, Padua, 1603. The valves are
shown at the points P, Q, R, and S.

Perhaps the best known work of Fabricius is that *On the
valves in the veins*. It had much influence on Harvey, who
borrowed figures from it and based much of his argument concerning the circulation of the blood on the action of these
valves. Fabricius' excellent figures of the valves in the veins are
the first in literature (Fig. 86). He explored them better than
anyone before his time, and they have often been regarded as his
discovery. Nevertheless, there can be no doubt that a number
of anatomists had already seen valves in the veins, and
justifiable claims may be made for the priority of Estienne,
Canano, Amatus Lusitanus, Vesalius, and Eustachius. This
does not remove the merit from the description of Fabricius.

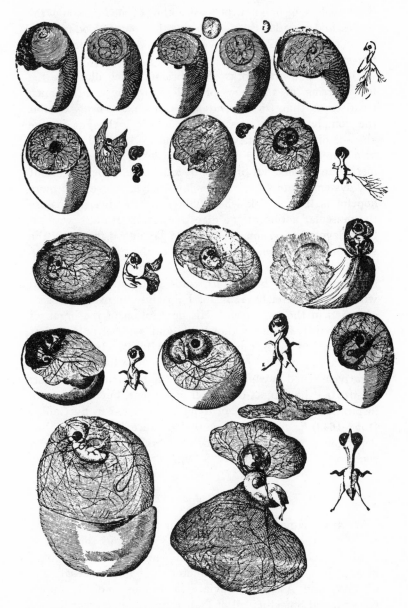

FIG. 87.—Page from Fabricius ab Aquapendente, *De formatione ovi et pulli*, Padua, 1600, showing early development of chick.

Nevertheless, he had not the least inkling of the function of the valves, and regarded them as slowing the flow of blood towards the periphery, and thus preventing blood from collecting at the extremities.

There are several works of Fabricius which illustrate the first stirring of the new physiological movement. Such treatises as that *On respiration and its instruments* exhibit the complete helplessness of physiological thought in the absence of any real knowledge of the workings of the heart or of the nature of the respiratory exchange. We have here merely an intellectual discontent with current views without any systematic building of new knowledge. Somewhat more hopeful is the outlook when Fabricius attempts to analyse the muscular action of the digestive tract. He also wrote a book devoted to vision, in which he gave good figures of the structure of the eye, being the first of the moderns to grasp the true form of the crystalline lens. The work is interesting in many respects, but is retrograde in others, as compared with that of Plater (p. 133), for Fabricius still places the seat of vision in the lens itself. His description of the organ of hearing hardly advances knowledge, and is no better than that of Coiter, but he is much happier in his treatment of the laryngeal apparatus. In dealing with it he adopts that comparative method in which we always see Fabricius at his best. He treated the subject of animal movement, but without the inspiration of the system of dynamics ushered in by Galileo (1554–1642) he could make no real advance. Success in that department was reserved for his more fortunately placed successor, G. A. Borelli (1608–79).

§ 9 *The last great Paduans, about* 1600–30

We are now nearing the end of our period, and may pause to survey the final scene. Almost to the end the lead in Anatomy remains with Padua. The new physiological movement is best represented by Sanctorius of Padua, Cesalpinus of Pisa, and Aselli of Pavia. Fabricius and his pupils Casserio and Spigelius, all of Padua, are the last great figures of the Vesalian line. After these men the intellectual hegemony

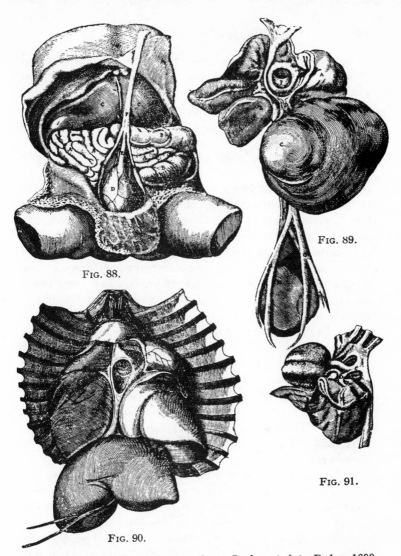

FIG. 88.

FIG. 89.

FIG. 90.

FIG. 91.

A page from Fabricius ab Aquapendente, *De formato fœtu*, Padua, 1600.

FIG. 88.—Abdomen of human fœtus opened. A, small intestines ; B, colon ; C, liver ; D, bladder ; E, urachus ; F, umbilical cord ; G, umbilical arteries.

FIG. 89.—Dissection of heart and neighbouring parts of a human fœtus. The legends run thus : " A, heart ; B, lungs ; C, liver ; D, vena cava passing from liver to heart ; E, right ventricle ; F, foramen in right ventricle with valve ; G, umbilical arteries ; H, bladder ; I, descending aortic trunk."

FIG. 90.—Shows same parts as Fig. 89, but in situ.

FIG. 91.—" Course of artery which passes from aorta to pulmonary artery." " A, heart ; B, lungs ; C, aorta ; D, end of vessel passing from aorta into pulmonary artery."

passes northward. Bauhin inaugurates the Paduan tradition at Basel, Bartholin and Wormius take it to Denmark, and Paauw to Holland. Riolan develops Anatomy at the isolated and independent university of Paris, and Harvey carries Paduan methods to England.

Santorio Santorio, or Sanctorius (1551–1636), is too nearly linked with the new physiological movement for us to do more than mention him in passing. That movement came to fruition in the next two generations and the real influence is to be seen in a later age than that which we are treating. The great achievement of Sanctorius was the introduction into Physiology of exact methods of measurement, pulse counting, temperature estimation, and weighing. Andrea Cesalpino (1619–03), professor at Pisa and Rome, rendered considerable service to botanical science. Italian writers have long urged his claims to the discovery of the circulation of the blood. In view of this, it seems right to mention him, though the claim seems to us quite untenable. It is true that there are certain passages in his writings quite consistent with the view that Cesalpino had attained to the conception of a circulatory movement of the blood. Unfortunately, however, in the opening chapter of his last work, the *Practice of the art of medicine* (1606), he makes, in fact, a formal statement of his belief that the blood *goes forth* from the heart not only through Aorta and Pulmonary Artery, but also *through Vena Cava* and *Pulmonary Vein* !

A physiological worker who comes more directly into our subject and period, however, was Gasparo Aselli (1581–1626). He was professor at Pavia, and in 1622 he discovered the lacteal vessels while dissecting a dog which had just been given a meal containing fat. These vessels had hardly been observed since Erasistratus (p. 32). The work of Aselli was published after his death. In it he sets forth his discovery in a sensational manner. Very large coloured plates illustrate this memoir, and it is the first book in which this device is adopted for anatomical purposes. The plates are extremely effective, more so than the garrulous and often difficult text (Fig. 92). They show the lacteals in animals, not in the human subject. The work was published the year before

Harvey's book *On the motion of the heart*, and it is evident that Harvey had not then seen it.

Giulio Casserio of Piacenza (1561–1616) was a pupil of Fabricius, whom he succeeded as professor at Padua in 1604. He greatly extended the knowledge of human Anatomy. Particularly he refined the Anatomy of the sense organs and of the organ of the laryngeal apparatus. In his works on these

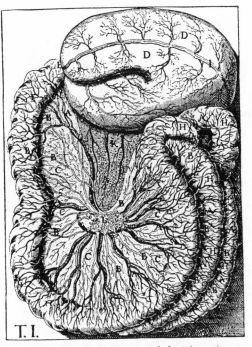

FIG. 92.—Gasparo Aselli, *De lactibus sen de lacteis venis . . . dissertatio*. This is not taken from the first edition, Milan, 1627, because the figures there are coloured and unsuitable for reproduction, but from the second edition, Basel, 1628.

structures he proceeds on most scientific principles. He first seeks to set forth a complete account of the organ in the human subject. Then he follows this organ through a long series of animal forms.

The method of Casserius is particularly well illustrated in his treatment of the apparatus of hearing. He describes and

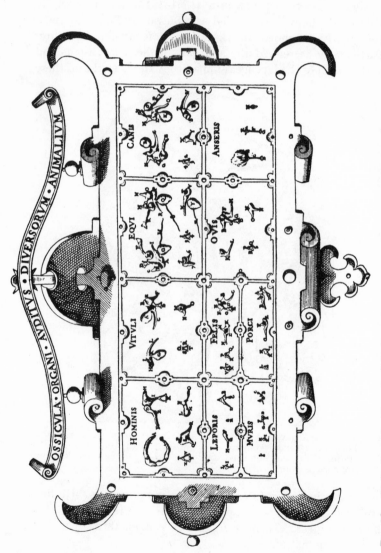

Fig. 93.—Giulio Casserio, *De vocis auditusque organis historia anatomica*, Ferrara, 1601. Ear ossicles of man, calf, horse, dog, hare, cat, goose, mouse, and pig.

figures the auditory structure of man, child, new-born infant, fœtus, ape, ox, horse, dog, hare, cat, sheep, goose, pig, mouse, turkey, and pike (Fig. 93). The investigation includes the ossicles and he shows careful dissections of the cartilage of auricle and external auditory muscles. Excellent too is his account of the vocal organs (Figs. 94 and 95). Copper plates are employed and these are capable of taking a much finer line than woodcuts. From his day and for another century and a half copper plate continued to be in almost universal use for purposes of anatomical illustration. Coiter, however, had used them to great effect as early as 1573, and Canano before that. Casserius set a very high standard both of workmanship and accuracy. His figures are the model for the copper plate illustrator as those of Vesalius and Ruini are for the woodcut operator.

On the death of Casserius in 1616 he was succeeded by his pupil Adriaan van der Spieghel (1578–1625), known as Spigelius. This man is the last of the great Vesalian line and, on the death of Spigelius, Padua ceased to lead the world in anatomical study. We observe that the last of the dynasty resembled the first in being a native of Brussels, and having studied first at Louvain. The anatomical works of Spieghel were not published till after his death, appearing in 1627. They contain a very large number of the plates of Casserius, which that author would have used had he lived (Figs. 96 and 97). The text of the work published in the name of Spigelius describes the lobe of the liver still called " Spigelian ". The volume contains many anatomical refinements, including the first adequate description of the spinal muscles which are well brought out in the series of very fine illustrations (Fig. 97). The chief interest in the book, however, is the great improvement it exhibits in anatomical terminology, which assumes a more convenient and a definitely more modern form, especially in connexion with the nomenclature of the muscles. This achievement was certainly the work of Spigelius himself.

With Spigelius what we may call the " heroic age " of Anatomy at Padua comes to an end. What were the reasons for its passing ? They appear to us to be two. On the one

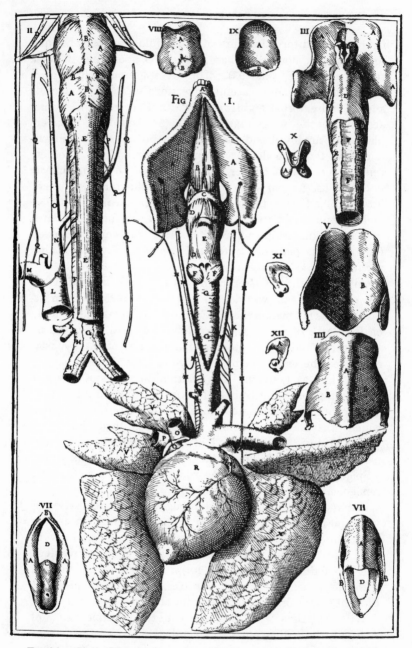

FIG. 94.—Giulio Casserio, *De vocis auditusque organis historia anatomica,*
Ferrara, 1601. Dissection of the vocal organs and parts relating
thereto of the pig.

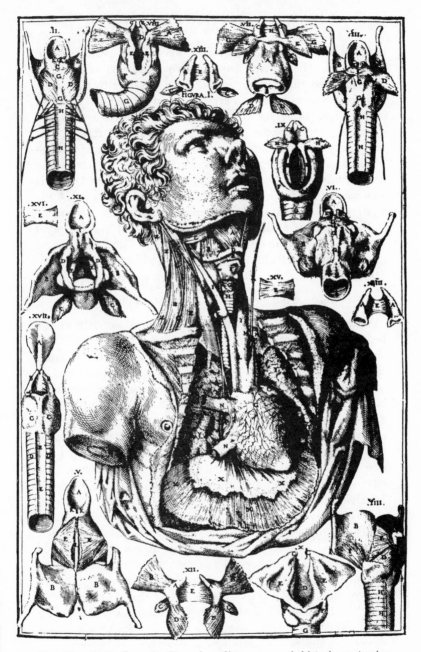

Fig. 95.—Giulio Casserio, *De vocis auditusque organis historia anatomica*, Ferrara, 1601. Dissection of vocal organs and parts relating thereto in man.

hand Spigelius worked exclusively on *human anatomy*. He abandoned the great comparative tradition that had distinguished the Paduan school from Vesalius to Casserius. His observations became more exact and refined, they gained in practical value for the surgeon, but they lost scientific interest.   On the other hand, with the work of certain contemporary investigators—Sanctorius, Van Helmont, Harvey—Physiology, rather than pure Anatomy, began to attract the best minds.   The new physiological era had opened, and it was some time before Padua attracted a physiologist of the front rank.   The star of Bologna rose again, and it was at Bologna, not at Padua, that Borelli (1608–79) and Malpighi (1628–94) worked in the next generation.   The children and grandchildren of Padua are to be sought in other lands than Italy.

§ 10   *Anatomy beyond the Alps*, 1590–1630

With the great development in anatomical activity at the beginning of the seventeenth century, the subject began to be studied more closely beyond the Alps.   In Switzerland Caspar Bauhin (1560–1624), professor at Basel, was distinguished as a Botanist, and in that capacity his name is still remembered. He erected an anatomical theatre in 1589.   In 1605 he produced a fine anatomical textbook which, though containing few original elements, was sound, scientific, and scholarly, and was often reprinted.   A few modern anatomical terms are due to Bauhin, among them *Areola* and *Phrenic nerve*.   He gave a good description of the muscles that move the tongue. In Holland dissection began at Amsterdam but developed chiefly at Leyden.   It was at the latter town that Peter Paauw (1564–1619), a pupil of Fabricius, and, like Bauhin, a professor both of Anatomy and Botany, built an anatomical theatre in 1597.   He published several anatomical treatises, annotated Vesalius, and made some additions to Anatomy, especially to that of the skull.   A pupil of Paauw was that Tulpius (1593–1674), who figures in the famous anatomical scene by Rembrandt (1632.   See also Plate XX).   At a somewhat later date great anatomical activity was developed at Leyden.

The Dane Olaus Wormius (1588–1654) studied under

Fabricius at Padua, and under Bauhin at Basel. He practised in London before settling at Copenhagen. He was a polyhistor and antiquarian, who devoted himself to many branches of learning, but is remembered by the " Wormian " bones. The term we owe to his more influential compatriot, Caspar Bartholin (1585–1629), who was the ancestor, in the flesh, of a regular dynasty of anatomists. After having been a student of Fabricius and Bauhin, Caspar Bartholin became Professor of Philosophy at Basel, of Anatomy at Naples, of Greek at Montpellier, and then successively of "Eloquence", of Medicine, and finally of Theology at Copenhagen. He spent his last few years in the study of Anatomy, on which he wrote tracts of little worth. They formed the basis of the work of his greater son Thomas (1516–80), whose life course lies outside our sphere.

In France practical Anatomy had been established for centuries at Montpellier, where we have to some extent followed it. An anatomical theatre was built there by Rondelet (p. 147) in 1556. At Paris, always the most conservative of Universities, the practice had long been known. We have followed there the work of Estienne and above all of Sylvius. Paris, however, had now become a humanist centre. For a time its scholars were more important than its dissectors. Much was done at Paris to help the anatomical Renaissance by the production of medical dictionaries. Of these the most important was the great Index to the works of Galen printed in 1550 by Antonio Musa Brassavola (1550–70), who, though an Italian, lived and worked in France. This fine work made Galen far more accessible and intelligible. It has never been excelled for completeness and accuracy, and is still in current use by modern scholars. I know no other work of scholarship of so early a date of which this can be said. A few years later, in 1564, there appeared at Paris two widely used philological dictionaries of medical terms. One was by Henri Estienne (1528–98), nephew of the anatomist Charles Estienne (p. 100). The other was by Jean de Gorris (1505–77). These three works defined and fixed a large number of anatomical words. They have exercised considerable influence on modern anatomical terminology.

An anatomist of small ability was Léonard Botal, who, though born in Italy, was of French parentage, and lived and worked in France. His very imperfect description of the *ductus arteriosus*, which we now know to be due to the persistence of the fifth cephalic aortic arch on the left side, appeared in 1565. To call the structure *ductus Botalli* is an anachronism, as it was in fact well known to Galen. After Sylvius there was little progress in Anatomy at Paris until the days of Jean Riolan the younger (1588–1657).

Riolan, like Bauhin and Paauw, was professor of both Botany and Anatomy. In the latter department he exhibited great activity, but a very conservative and yet controversial spirit. He opposed, for instance, the teaching of Harvey. He was, however, an extremely popular teacher, and made many additions to anatomical knowledge ; thus he gave a good description of the mesentery ; he described the *appendices epiploicæ* ; he improved the knowledge of the spermatic vessels, and discovered the embryonic gill slits. Riolan, owing to his popularity as a teacher, was able to introduce many words into Anatomy ; among them are *Deltoid, Pectineus, Scalenus*, the names that end in *-glossus*, e.g. *Hyoglossus*, together with certain terms which reinterpret phrases of Galen, e.g. *Thalamus opticus* and *Arytenoid*. His concise style of writing made his textbooks very popular with students. The most widely read anatomical works of the seventeenth century bear the names of Riolan and Bartholin.

In Germany and Austria, countries then included in the so-called Holy Roman Empire, but little progress was made during our period. The *Fabrica* of Vesalius was brought out in a German edition in 1543, but there were few later original contributions. To Dryander we have already referred (p. 98). Coiter worked at Nuremberg. At the very end of our period that admirable surgeon, Hans Faber of Hilden, known as Fabricius Hildanus (1560–1634), made minor contributions. The fact that dissection had been practised at a number of the Universities of the Holy Roman Empire, from the Middle Ages onward, is in itself enough to prove that something more was wanted for anatomical progress. That something, as we have seen, was

Fig. 96.—Plate prepared by Casserius, but first published in Adrian
Spigelius, *De formato fœtu*, Padua, without date, but with
dedication of 1626.

the combination of Renaissance Art and Humanist Learning
with enthusiasm for dissection.   Renaissance Art and
Humanist Learning were somewhat late in coming to German
lands.   It was after our period that they became united
with Anatomy, and there was then a movement similar
to that which we have seen elsewhere.

§ 11   *The Beginnings of Anatomical Study in England* (1360–1640)

(*a*) *The Middle Ages.*

The first work of anatomical import to appear in the English
language was the *Surgery* of Lanfranchi of Milan (died about
1306), a Bologna student and a pupil of William of Saliceto.
The English translation of Lanfranchi contains several sections
on Anatomy which resemble those of his master.   The English
version was made about 1380, and the work of William of
Saliceto soon followed.   Mondeville's *Surgery* was rendered
into English in the fifteenth century.   More popular than any
of these was a translation of the anatomical section of Guy de
Chauliac, which, though never printed, has survived in a
number of fifteenth century manuscripts.   Several other
anonymous tracts on Anatomy based on those we have
mentioned also appeared in English in the fourteenth and
fifteenth centuries.   Of the fifteenth century is also a Latin
manuscript of John of Arderne (1307–*c.* 1380) which contains crude anatomical figures.

Printing began in London toward the end of the fifteenth
century.   In 1495 Wynkyn de Worde printed there the
Encyclopædia of Bartholomæus Anglicus (flourished 1260),
containing a fine illustration representing a dissection scene
(Fig. 98).   The first printed book actually devoted to Anatomy
in English carried the name of a Barber-Surgeon, Thomas
Vicary (died 1561) and the date 1548.   It was, however, not
original, but simply copied from a fourteenth century English
manuscript based on Lanfranchi and Mondeville.   Nevertheless, it remained long in use, and was reprinted in 1577 by
the surgeons of St. Bartholomew's Hospital and appeared in
a number of later editions.   It is a purely mediæval work.

PLATE XVI

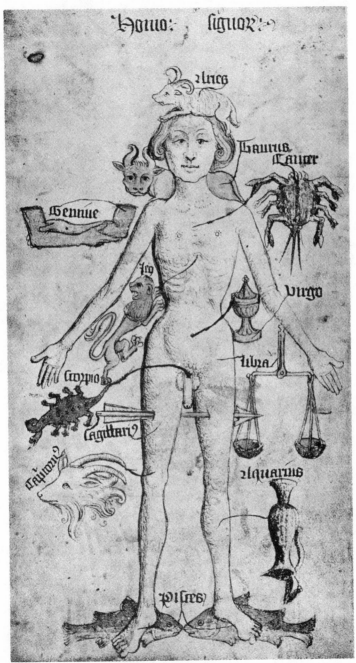

ZODIACAL MAN (about 1450)

from the Guild Book of the Barber Surgeons of York now in the British
Museum. The signs of the Zodiac are written on the parts of the body
which they are supposed to influence, from *Aries*, the ram on the head,
to *Pisces*, the fishes on the feet.

[*face p. 170*]

(b) *The Renaissance.*

The Renaissance alike of Letters, of Science, and of Art came late to England. The earliest to bring the medical humanism to this country was Linacre (p. 105), with his translations of Galen and his foundation of the Royal College of Physicians in 1518. An Englishman, Michael Hatchett, studied under Sylvius in Paris in 1536, and seems to have been in touch with Vesalius, then in his earlier Galenic phase. Securis, who afterwards returned to England, bought some of the books of Vesalius when the great anatomist left Paris.

John Caius (1510–73) visited Padua as early as 1539. He came in contact with the Humanist Montanus (p. 105), and spent much time in the study of Greek and actually resided for eight months in the house of Vesalius himself. He then travelled widely with the object of obtaining good manuscripts of Galen and Hippocrates. He returned to England in 1544, and began to give lectures in London on Anatomy. These he continued for twenty years. He edited some of the anatomical works of Galen. Caius was a confirmed and obstinate Galenist of the old school, and added nothing to anatomical knowledge.

The work of Vesalius was pirated in England by one Geminus in 1545. This book exerted considerable influence. It is the first book printed in English for which copper plates were used. Three editions of it appeared. The second, issued in 1553, contains a translation into English by the infamous Nicholas Udall the dramatist, author of " Ralph Roister Doister ", and successively headmaster of Eton and Westminster. The third edition of 1559 contains a portrait of Queen Elisabeth on the title page which is said to be the first engraved of Her Majesty.

In 1540 Henry VIII licensed the Barber Surgeons to anatomize the bodies of four felons a year. Demonstrations were given at the Hall Company in London. In 1557 attendance at them was compulsory to members. An anatomy theatre was erected by the Company. It had to be enlarged in 1568. For a few years around 1566 demonstrations on Anatomy were given at the Barber Surgeons' Hall by Giulio Borgarucci, an Italian refugee, who had perhaps been

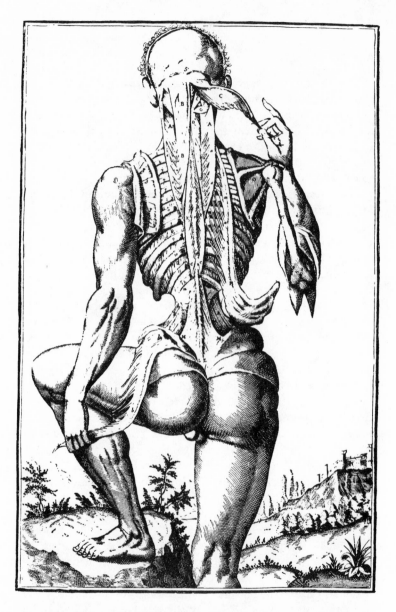

Fig. 97.—Plate probably drawn for Spigelius, but appearing first in a work bearing the title *Julii Casserii Placentini, Tabulæ anatomicæ LXXIIX, omnes novæ nec ante hac visæ; Dan. Bucretius XX quæ deerant supplevit et omnium explicationes addidit*, Venice, 1627. It exhibits an attempt to analyse the spinal muscles.

a student at Padua, and whose brother professed Anatomy at that University. Borgarucci, however, added nothing to knowledge. In 1565 Elizabeth granted the right to dissect to the College of Physicians.

In 1572 John Bannister (1533–1610) was admitted a member of the Barber Surgeons' Company, and was soon after

FIG. 98.—Dissection scene from the English translation of Bartholomæus Anglicus printed by Wynkyn de Worde at London in 1495. This is the first picture of dissection in a book printed in England.

appointed their Anatomical Lecturer. In 1578 he published in English a work on Anatomy containing no original elements. A picture of him delivering the visceral lecture in 1581 at the Barber Surgeons' Hall is well known. He stands by the side of an open copy of the work of Columbus. We have also a

number of anatomical drawings and models by Bannister.
These, with few exceptions, are all copied from Vesalius,
probably via Geminus. Among the exceptions are a small carved
ivory model of a skeleton now in the Library of the University
of Cambridge, and a sheet of drawings of the skeleton now in
the Hunterian Library at Glasgow (Plate XVII). They are
both inaccurate, but worthy of attention as the first indications
of independent anatomical investigation in this country.

The general barrenness of anatomical study in England
before Harvey is well illustrated by the *Anatomy Lectures*
of Thomas Winston (1575–1655), which were delivered by
him at Gresham College during Harvey's most active years
and printed posthumously. All Winston's Physiology and
nearly all his Anatomy is still purely Galenic.

The state of English Anatomy was utterly changed by the
advent of William Harvey (1578–1657). He spent the
years 1593–7 at Caius College, Cambridge, where he may
well have been infected by the Paduan tradition established
there by John Caius. Harvey was at Padua from
1597 to 1602 when Fabricius was at the height of his
powers. Returning to England full of the comparative
method which Fabricius was expounding, he put it to good
account. His first lectures, the result of years of thought and
experiment, were given at the Royal College of Physicians.
in 1615–16. His notes for these lectures are now in the
British Museum, and show that he had already grasped the
idea of the circulation of the blood. It was another twelve
years before he allowed himself to appear in print.

§ 12 *The Work of William Harvey*, 1628

Harvey's great work, *An anatomical dissertation on the
movement of the heart and blood in animals*, appeared at
Frankfort in 1628, as a miserably printed little quarto. By
this brief tract—for it is little more—the whole scientific
outlook on the human body was transformed. From now
on, men begin to think *physiologically* even when occupied
in purely anatomical study. Harvey took up his theme

PLATE XVII

[face p. 174

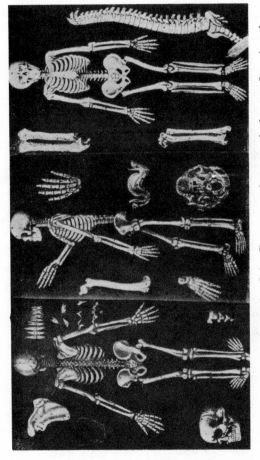

Figures of a skeleton, used by John Banester and prepared about 1580. A volume containing this and other anatomical figures by him is now in the Hunterian Library at Glasgow. The figures here reproduced are probably the earliest prepared in this country which were drawn from the object. Earlier English figures of the skeleton are known, including an ivory statue that was in Banester's own possession. These, however, are all copied from earlier works.

practically where Galen had left it. With Harvey, at last, a clear idea again emerges that each organ has a discoverable function and is related in its mode of working to all the other organs and to the body as a whole. The point of view of Harvey, however, is very different from that of Galen and in the coming centuries we hear less of *Design* and more of the *Machine*.

Apart from the superb tenacity and experimental skill with which he follows his great theme, Harvey's main scientific virtue is a negative one, though it is none the less a virtue on that account. It is the virtue of restraint. He refuses to discuss more remote problems until those nearer at hand can be solved. He thus declines the everlasting debate on such topics as the nature of life or the origin of the innate heat. His self restraint is the more remarkable when we remember that he was in fact an extreme and convinced Aristotelian. These questions concerning the nature of life and animal heat had employed active minds and fluent pens for centuries, and yet no progress had been made toward their resolution. From now on, fired by the example of Harvey, investigators begin to occupy themselves with the interpretation of the problems of Anatomy and Physiology in the more comprehensible terms of Mechanics, of Physics, of Chemistry and of Comparative Anatomy.

The book which started this movement, though very convincing, is by no means easy reading. We have observed that there are special difficulties in the interpretation of the three previous writers, who provide the great landmarks in the History of Anatomy, Galen, Mondino, Vesalius. The difficulties of Harvey's book are, however, very different from those of the *Fabrica* of Vesalius, the *Anathomia* of Mondino, and the *Anatomical Procedure* and *Uses of the bodily parts of man* of Galen. It is with Vesalius that one is most tempted to contrast Harvey. The two men are near to each other in time. Both are connected with Padua. Both are part of the same intellectual movement. Yet their two minds are poles asunder. Harvey is extremely conservative, a philosopher by temper, cautious, slow, devoid of literary or oratorical charm or gift. His forces are not diffused in unnecessary learning, but he is very tenacious of positive knowledge. Vesalius is a radical, a man of lightning quickness of mind, intent on an immediate end and that a practical one, full of the eloquence native to exuberent spirits and forceful strength, learned and, perhaps, above all things, an artist with some of the defects of an artist's qualities.

The modest, reticent Harvey makes none of that display of erudition of which Vesalius was so fond. The work of Harvey, however, like that of Vesalius, frequently refers to ancient authorities with whom the modern reader is unfamiliar. It was necessary for him to deal with these older writers, because they provided the physiological views current in the seventeenth century. To a reader nowadays, however, the constant return to the opinion of the ancients forms an obstacle to the understanding of his argument. We may remember, too, that in Harvey's time, men were unaccustomed to the physiological method of research. He had, therefore, to devote a considerable portion of his little space to justify the process of applying to one creature the deductions drawn from another, matters that are now mere physiological commonplaces.

Harvey's work, however, contains other obstacles to comprehension for which, perhaps, less justification can be found. Thus the Latin style selected by him is peculiarly ill-adapted to scientific discussion. The sentences are often exceedingly involved, usually unduly long, not always of clear construction. His division into chapters does not really correspond to the natural division of the subject. Harvey, too, was a very careless proof corrector—if indeed he ever looked at his proofs at all—and thus an intolerable number of printer's errors have survived in the text. Lastly, there are certain practical reasons that make his book difficult. One is that it is extremely condensed and contains a vast number of conclusions and observations packed into a very small space. Another is that it was the first time that a treatise had been devoted to a physiological theme and Harvey is grappling with what is, in effect, the rebirth of a method that had been neglected for nearly 1,500 years. In summarizing Harvey's argument, it will therefore be best to give his conclusions mainly in our own words, not in his. It will conduce to clarity if we reproduce the steps in his reasoning in a series of sections which only partially correspond to the chapters in his great book.

Before we actually describe his observations, experiments and inferences, we may consider the knowledge of the subject available in his day. The general structure of the heart had been well known since Vesalius. It had been accurately studied by several of the successors of Vesalius. The action of the valves in the Aorta and the Pulmonary Artery in preventing regurgitation of blood had been described by Galen and recognized by Mondino, Leonardo, Berengar, Vesalius, and several later writers. The septum had been regarded, however, as perforate, and it was considered that blood passed through it from the right ventricle to the left (Fig. 30). This view of Galen, accepted by Mondino and his successors, and even by Leonardo (Plate XI) had been treated with scepticism by Vesalius (p. 132) who, however, had nothing better to put in its place. The lesser circulation had, however, been described by Servetus, Columbus, Ruinus and others, and must have been fairly well known though its importance was quite unrecognized. Cesalpino had given a hint of the greater circulation, but had withdrawn it (see p. 160). Finally, the valves in the veins, seen by many during the sixteenth century, had been systematically explored by Fabricius, who had no idea of their real action.

We may now turn to the work of Harvey, reproducing his reasoning in a series of steps.

(a) Preliminary observations

1. The heart, if grasped, is felt to become harder during action. This hardness proceeds from the same kind of tension as that of the muscles of the forearm during contraction. The contraction or *systole* of the heart therefore corresponds to its *active* position. The action of the heart, in fact, must be considered like that of any other muscle.

2. In cold-blooded animals, especially, it may be observed that when the ventricles contract, they become lighter in

colour, and that when they expand they become darker in colour. This point had been observed by many earlier workers, notably, by Coiter with whose writings Harvey was familiar.

3. During action the heart is observed to become erect, hard, and of diminished size. During contraction, moreover, while the size and breadth of the heart decrease, the length actually increases. Thus it is that the apex of the heart strikes the chest wall during contraction. As we have seen, these points were not wholly unknown to previous writers, Coiter among them (p. 149), but no special importance had been attached to them.

4. The contraction of the heart and the contact of its apex with the chest wall are *simultaneous with the expansion of the arteries*, as felt at the pulse. This most important observation disposed of the older view that the dilation of the artery was the result of its active expansion. It also was known to Coiter, who failed to perceive its significance.

5. The previous observation made it highly probable that the *contraction of the heart was the cause of the expansion* of the artery. The conclusion was confirmed by the character of the bleeding from a cut artery, for the spurt can be seen to take place at the moment of ventricular contraction, and not during ventricular expansion. In this simple interpretation of a known fact, Harvey seems to have been a pioneer. The conclusion was further confirmed in a letter to Riolan (p. 168) by an observation which Harvey made on a case of calcification of an artery. Harvey observed pulsation during life below the area of calcification and confirmed the calcification by postmortem examination.

6. In a series of experiments the auricles were shown to have somewhat similar relations to the ventricles as the ventricles have to the arteries. Thus, if the tip of the ventricle be removed, each beat of the auricle is seen to be followed by a corresponding spurt of blood from the open cavity of the ventricle. In the dying heart the ventricles cease beating before the auricles. This observation of the relation of auricles and ventricles seems to be wholly Harvey's own.

7. The contraction of the auricle is *followed* by that of the

ventricle. The same blood, therefore, that is driven into the ventricle by the contraction of the auricle is subsequently driven into the arteries by the contraction of the ventricles. The point is Harvey's own.

8. Once the blood has entered one of the great arteries—whether the aorta or the pulmonary artery—it cannot come back along the same path. Its return is absolutely stopped by valves. This observation, made by Galen, had been repeated for both the great vessels by Leonardo. Leonardo's writings were not available to Harvey. The action of the valves of the pulmonary artery had, however, been recognized by a series of writers from the time of Servetus and Columbus. The conclusions of Columbus were familiar to Harvey.

9. Although the action of the cardiac valves had been thus recognized by many previous writers, Harvey introduces a new point in this connexion. The point is now so perfectly obvious that it seems extraordinary that no previous writer had made it clearly. Harvey insists that the flow of blood is not only in one direction, but is *continuously* so. This leads him to a very crucial discussion. Consider, he says, the capacity of the heart. Suppose the ventricle holds but two ounces. If the pulse beats seventy-two times in a minute, in one hour the left ventricle will throw into the aorta no less than $72 \times 60 \times 2 = 8640$ ozs. $= 38$ stone 8 lb. i.e., three times the weight of a heavy man ! Where can all this blood come from ? Where can it all go to ? The point is made with great force and I know no parallel to it in earlier literature.

10. It can only be from the veins that all this blood must come which is sent out continuously by the aorta. This conclusion was reinforced by a very simple experience. If an artery is cut, the animal bleeds to death. The bleeding could be seen to get slower and slower until finally it ceased as the blood was exhausted and death approached. The reason must be that the blood being lost does not reach the veins, and so cannot return to the arteries. Here, again, Harvey puts his own true and simple interpretation on a perfectly well-known fact.

(b) The Solution

11. But how does this blood get from the venous system to the left side of the heart and how does it get from the

arteries into the veins ?   To these crucial questions we now know the answers. Let us see how Harvey answered them.

Blood, he observed, can enter the right auricle through the vena cava, the opening of which is patent. It can then enter the right ventricle, the opening into which from the right auricle is equally obvious though guarded by valves which prevent its regurgitation. From the right ventricle, therefore, Harvey argued, there can be but one exit, the *pulmonary artery* (or as Harvey called it the arterial vein). If it enter that vessel it cannot again return, as he well knew, for he was well acquainted with the action of the sigmoid valves (see above, under 8).

We may here digress for a moment to consider Harvey's own mental state at this point. The whole of his book is saturated with the ideas of Galen and Aristotle. Harvey never succeeded in freeing himself from the hypnotism induced by these authors. Thus at this very stage, he exclaims " from Galen, that divine man, that Father of Physicians, it *clearly appears* that the blood passes through the lungs from the arterial vein (*pulmonary artery*) to the small branches of the venal artery (*pulmonary vein*) ". Now Harvey was certainly wrong here. The passages in the work of Galen that refer to this matter are, in fact, neither clear nor are they consistent one with the other. There is, however, just one short paragraph in Galen's very voluminous works which may be given the interpretation that Harvey puts on it. It is a strange thing that *at the moment of his making the discovery on which his fame rests, Harvey should ascribe it to another* ! This attitude must appear fantastic to many modern readers. To explain it would demand an investigation of the general question of the psychology of scientific discovery. Here we can but remind our readers that Harvey rather prided himself on being extremely conservative in his mental outlook.

12. The last step in the solution of the problem that Harvey had now before him is best stated in his own words. " I began to think whether there might not be *a movement, as it were, in a circle,* and this I afterward found to be the case. I saw that the blood, forced by the action of the left ventricle into the artery, was sent out to the body at large. In like manner

PLATE XVIII

Gulielmus · Harvey · M·D·

WILLIAM HARVEY 1578-16
From the painting by CORNELIUS JANSSEN in the Royal College of
Physicians of London.

it is sent to the lungs, impelled there by the right ventricle into the arterial vein (*pulmonary artery*)."

He had, in fact, begun to think this at least as early as the year 1615. This is proved by remarks in the notebooks of his lectures delivered that year. That he had withheld publication so long argues extraordinary caution, patience, and self-restraint.

13. Having reached this point, the explanation of other matters soon followed. Thus it was now manifest why in dissection of dead bodies there is so much blood in the veins and so little in the arteries, so much in the right ventricle, so little in the left. " The true cause is that there is no passage to the arteries save through lungs and heart. When an animal ceases to breathe and the lungs to move, the blood in the arterial vein (*pulmonary artery*) no longer passes therefrom into the venal artery (*pulmonary vein*) and thence to the left ventricle. But the heart not ceasing to act as soon as the lungs, but surviving them and continuing to pulsate for a while, the left ventricle and the arteries continue to distribute their blood to the body at large, and to send it into the veins."

(c) Some crucial experiments

14. Harvey now proceeds to examine serpents in which the great vessels are very conveniently arranged for observation. When one of these animals is laid open, the vena cava may be compressed by means of forceps. The part between the forceps and the heart then becomes empty almost immediately. The heart, too, will become pale, even when dilated. It also becomes smaller since it is no longer filled with blood. At last it begins to beat more slowly as if about to die. If now the impediment to the flow of blood be removed, the colour and size of the heart are instantly restored.

15. The same experiment is now performed on the aorta. He observes that the part between the obstruction and the heart itself became inordinately extended, as though about to burst, and it assumes a deep purple colour. On removal of the obstruction, colour, size, and pulse return at once.

16. Harvey now makes some simple experiments on the

arm of the living man (Fig. 99). By bandaging above the elbow, he was able to compress either the vein alone or arteries together with veins and to observe the various effects of the two different kinds of pressure. Moreover by bandaging so as to obstruct the veins alone he was able to bring out knots in them. By placing the fingers of one or both hands along the veins at different points, he was able easily to show that the flow of blood took place toward the knots and not away from them.

The interpretation of the knots themselves was familiar to Harvey. He knew that they correspond to the valves which had already been illustrated by his teacher Fabricius (Fig. 86) and seen by various earlier workers (p. 156). Harvey, in fact, borrows his figure from the work *On the valves of the veins* by Fabricius.

(d) Conclusion

17 " All things, both argument and ocular demonstration, thus confirm that the blood passes through lungs and heart by the force of the ventricles, and is driven thence and sent forth to all parts of the body. There it makes its way into the veins and pores of the flesh. It flows by the veins everywhere from the circumference to the centre, from the lesser to the greater veins, and by them is discharged into the vena cava and finally into the right auricle of the heart. [The blood is sent] in such quantity, in one direction, by the arteries, in the other direction by the veins, as cannot possibly be supplied by the ingested food. It is therefore necessary to conclude that the blood in the animals is impelled in a circle, and is in a state of ceaseless movement ; that this is the act or function of the heart, which it performs by means of its pulse ; and that it is the sole and only end of movement and pulse of the heart."

The last paragraph may well be committed to memory as being the foundation on which our conception of the workings of the animal body rests. In the old physiology the arteries were considered as distributing spirit and higher forms of vital activity, while the veins distributed nourishment and the lower forms of vital activity. The arteries arose from the heart ; the veins, it was thought, from the liver.

In the Galenic physiological system (Fig. 30) the right

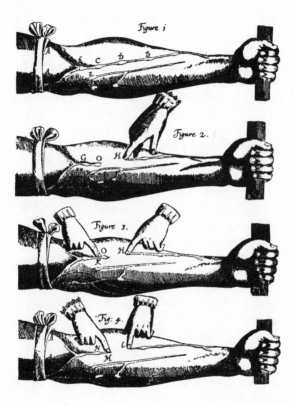

FIG. 99.—Experiments on bandaged arm from William Harvey, *Exercitatis anatomica de motu cordis*, Frankfurt, 1628.

ventricle was, in effect, one of the main branches of the venous system. The vessel, therefore, that arose from the right ventricle and linked it with the lung was thus itself a branch of the same system. It was therefore a " vein "

As it had thick walls like an artery it was called the *arterial vein*. It is the vessel we now call the *pulmonary artery*.

A companion to this vessel connected also the *left* cavity of the heart with the lung. As a derivative of the left side of the heart from which the aorta arose, it was regarded as an " artery ". Its walls, however, were thin like those of a vein, and it was called, therefore, the *venal artery*. It is the vessel we now call the *pulmonary vein*.

A result of the discovery of Harvey was the introduction of a new nomenclature. For us who look at the circulation from his point of view, new names are needed. The arterial vein has become the pulmonary artery, and the venal artery has become the pulmonary vein. For us an artery is a vessel which takes blood *from* the heart, and a vein is a vessel which takes blood *to* the heart. Until Harvey's day, arteries were distinguished from veins by containing a different kind of blood. Men had no idea of the constant and massive change of arterial into venous, and of venous into arterial blood. Our conception of the difference between artery and vein is controlled by our idea of the circulation introduced by Harvey. A new nomenclature was therefore needed and it came with the new age which Harvey ushered in. It is a change that Harvey himself foresaw.

Before we quite part with Harvey, we would note that that great discoverer did not use a microscope. He leaves open the question as to whether the blood in passing from veins into arteries is retained in vessels or passes into pores and cavities in the tissues. The use of the microscope—invented about 1608 — profoundly affected the development of Anatomical thought in that new era of which Harvey had a Pisgah sight.

§ 13 *Epilogue*

From Harvey's time onward the tradition of Padua has never departed from our medical schools. It is interesting to remember that the tradition has since been reinforced, for the first occupant of the first Chair of Physiology established in England was himself a student of Padua. That occupant

Plate XX

Dissection scene painted by Rembrandt (1606-1669) in 1656. Amsterdam. The picture has been burnt but this central part remains. It is influenced by Mantegna, see Plate XIII.

[*face p. 184*

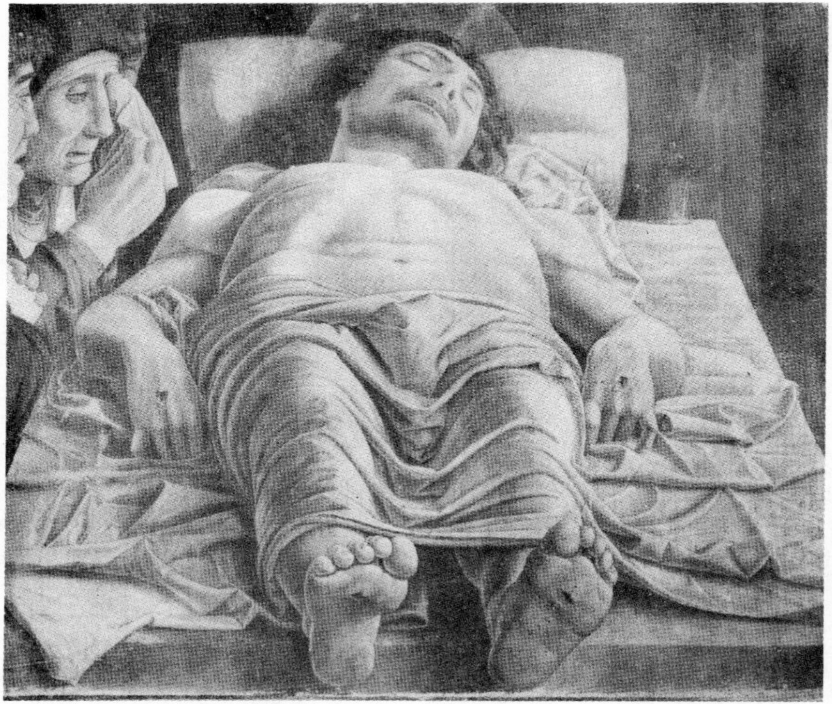

Plate XIX

ANDREA MANTEGNA (1431-1506), THE DEAD CHRIST

This picture, which is in the Brera Palace at Milan, is a careful anatomical study from nature. It was imitated by Rembrandt (see Plate XX.).

[face p. 185

was William Sharpey (1802–80), and the Chair was at University College, London. Sharpey's own copies of Vesalius, of Fallopius, of Columbus, of Casserius, of Spigelius, of Harvey, and of other Paduan anatomists lie open before me as I write. They bear upon them the evidence of Sharpey's vivid consciousness of the antiquity and dignity of the line from which he was descended.

It is not unfitting to close with a reflection as to the present state of Anatomy in England. As a result of the rapid accumulation of knowledge during the last century the departments of anatomical study have been subdivided. The subdivision has unfortunately been along the lines of supposed ' practical ' needs. It has had no regard to that natural specialization which depends on the faculties and endowments of the human mind rather than on the mere mass of facts accumulated in any particular department. Subjects such as Anatomy, Physiology, and Histology have been arbitrarily separated, and have come to be regarded as but distantly related to each other. This unhealthy specialization and separation has been encouraged by the necessities, or supposed necessities, of medical education as developed towards the end of the century. The rigid separation of these studies has been an evil to them all, but Anatomy perhaps has suffered most. It is therefore appropriate that the Institution to which Sharpey, the last of the English Paduans, devoted his life, should be the first to take organized steps to reunite these studies and to bring them into contact also with Comparative Anatomy and Embryology. In doing this University College has returned to the great Paduan tradition as it existed from Vesalius to Harvey.

A VESALIAN ATLAS

CONTAINING

NUDES, SKELETONS AND MUSCLE TABULÆ
FROM THE *EPITOME* AND *FABRICA*

FIG. 100.—Nude Exhibiting the Canon of Proportion, from
the *Epitome*.

Fig. 101.—Nude Exhibiting the Canon of Proportion, from the *Epitome*.

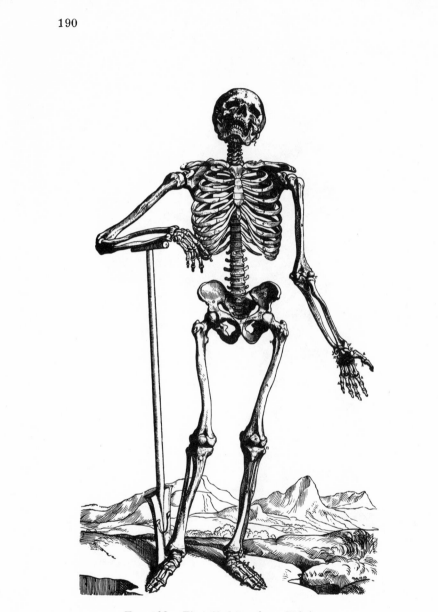

Fig. 102.—First Skeleton from *Fabrica*.

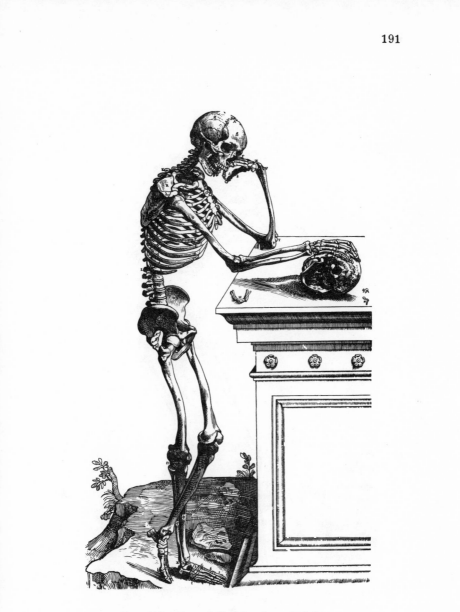

Fig. 103.—Second Skeleton from *Fabrica*.

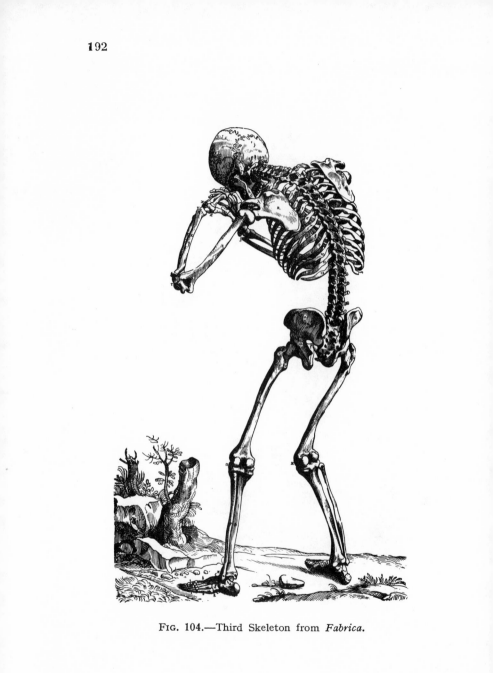

FIG. 104.—Third Skeleton from *Fabrica*.

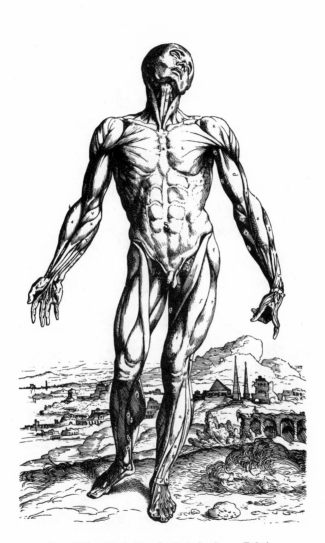

FIG. 105.—First Muscle Tabula from *Fabrica*.

Fig. 106.—Second Muscle Tabula from *Fabrica*.

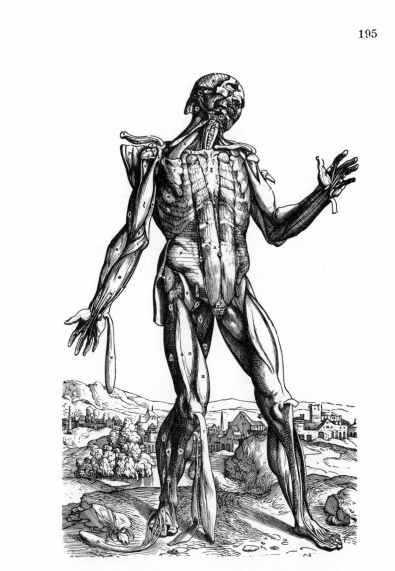

FIG. 107.—The Fifth Muscle Tabula from *Fabrica.*

Note extension upward of *rectus abdominis* as in apes, and the muscle in the neck marked X which has no existence in man but is to be found in apes. Vesalius calls attention to these points in the text.

196

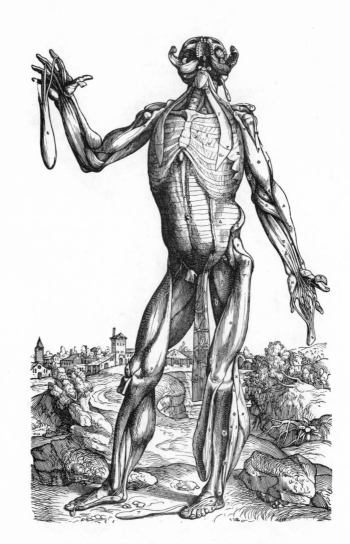

FIG. 108.—Sixth Muscle Tabula from *Fabrica*.

Note scalene muscle continued as a strip in front of ribs anterior to the *serratus magnus*. The description is drawn, as Vesalius points out in the text, from the dog.

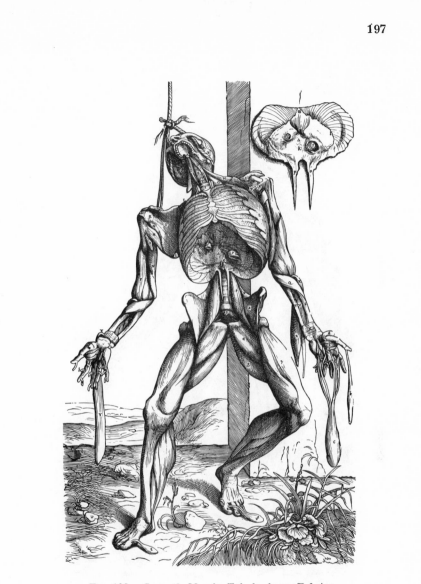

FIG. 109.—Seventh Muscle Tabula from *Fabrica*.

For pose compare figure on Pl. XI. The diaphragm is
shown separately above.

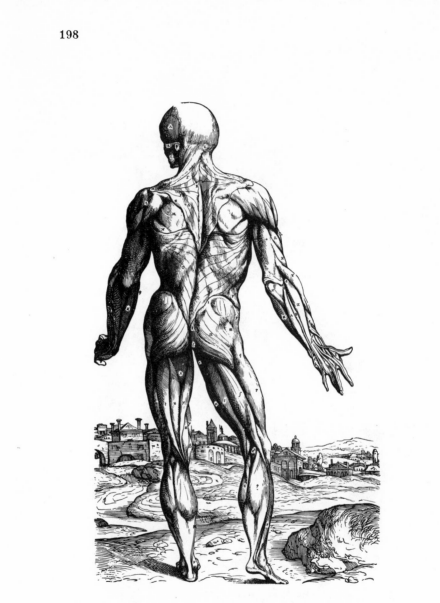

Fig. 110.—The Ninth Muscle Tabula from *Fabrica*.

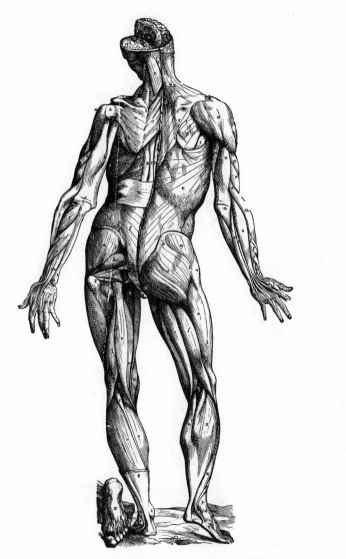

Fig. 114.—Fourth Muscle Tabula from *Epitome*.

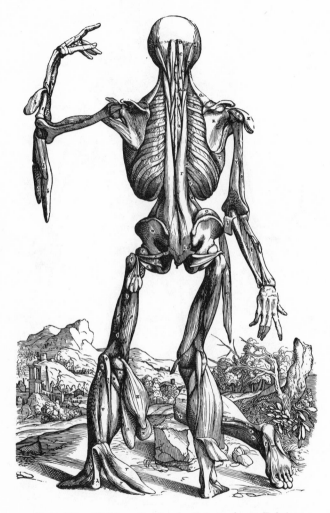

Fig. 111.—Thirteenth Muscle Tabula from *Fabrica*.

200

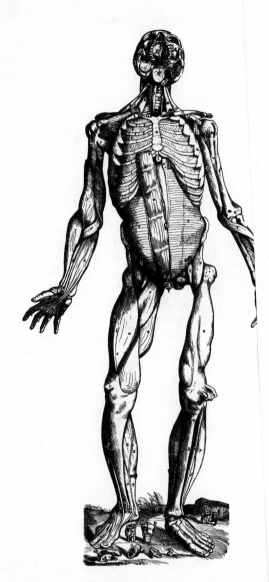

FIG. 112.—Fifth Muscle Tabula from *Epitome*.

In this and the following figures the right side of the dissections exhibits the more superficial and the left the deeper muscles.

FIG. 113.—Third Muscle Tabula from *Epitome*.

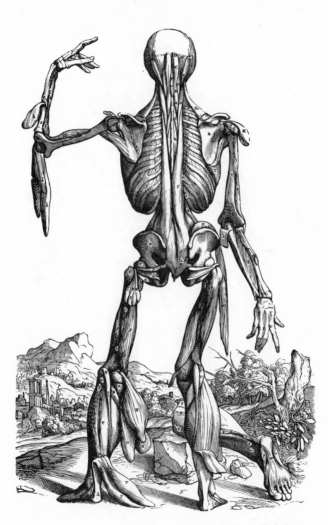

FIG. 111.—Thirteenth Muscle Tabula from *Fabrica.*

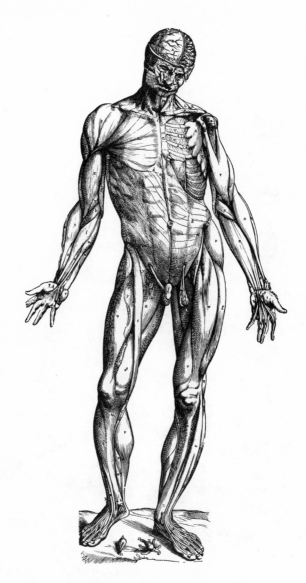

FIG. 112.—Fifth Muscle Tabula from *Epitome*.

In this and the following figures the right side of the dissections
exhibits the more superficial and the left the deeper muscles.

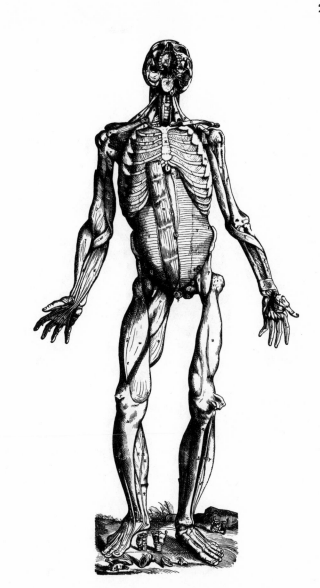

Fig. 113.—Third Muscle Tabula from *Epitome*.

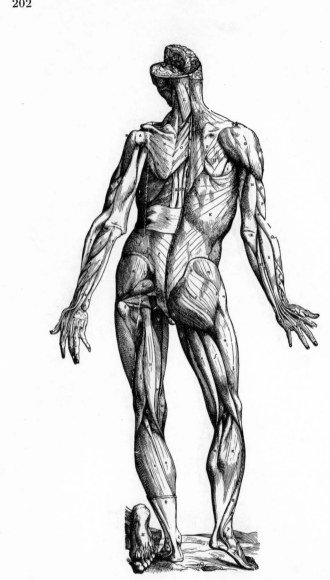

Fig. 114.—Fourth Muscle Tabula from *Epitome*.

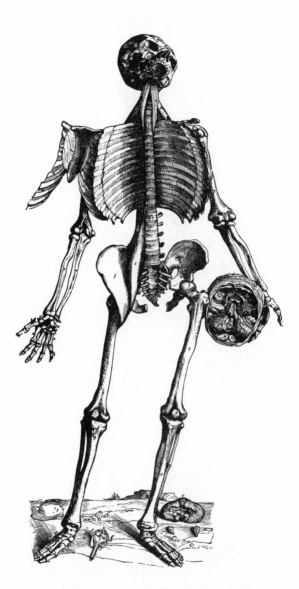

FIG. 115.—First Muscle Tabula from *Epitome*.

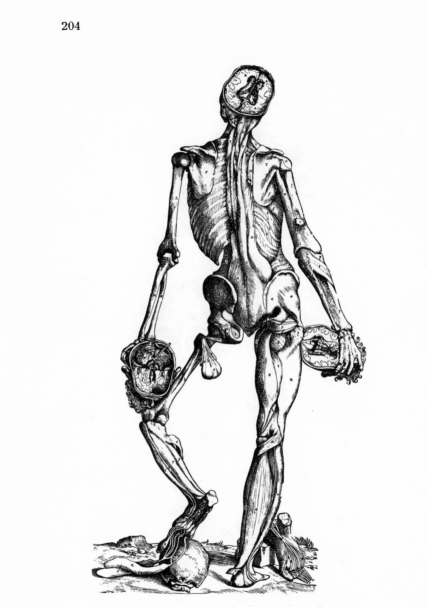

FIG. 116.—Second Muscle Tabula from *Epitome*.

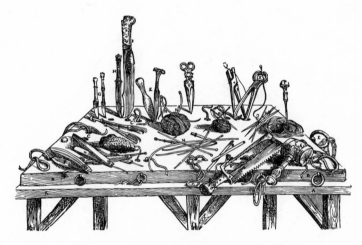

QVADRAGESIMIPRIMI CAPITIS FIGV-
rarum,eiufdemíq characterum Index.

FIG. 117.—The Instruments used by Vesalius, from the *Fabrica*.

INDEX OF PERSONAL NAMES

Rabelais, 147.
Raphael, 90.
Reil, 115.
Rembrandt, 166.
Remmelin, 128.
Rhazes, 68, *et passim.*
Riolan, 160, 168.
Rondelet, 147 f., 167.
Rufus, 42 ff.
Ruinus, 153, 178.

Sanctorius, 158, 160, 166.
Satyrus, 46.
Salimbene of Parma, 72.
Senac, 101.
Servetus, 140, 142, 178.
Severus, Alexander, Emperor, 37.
Sharpey, 185.
Signorelli, 90.
Sixtus IV, Pope, 86, 121.
Socrates, 15.
Soranus, 44 f.
Spigelius, 57, 115, 161 ff., 185.
Stephen of Antioch, 68.
Swammerdam, 32.
Sylvius, 57, 107–9, 117, *et passim* throughout.

Theophrastus, 22.
Tertullian, 34.
Thaddeus of Florence, 72, *et passim.*
Theodoric, Emperor, 37.
Theodoric of Borgognoni, 71.
Tulpius, 166.

Udall, 171.

Valsalva, 115.
Varolio, 143 f.
Verocchio, 90.
Vesalius, 111, *et passim* throughout.
Vespasian, Emperor, 37.
Vicary, 170.
Vidius, 34, 144.

Wharton, 89.
William of Saliceto, 71, 72, *et passim*, 170.
Willis, 115.
Winston, 174.
Worm, 160, 166 f.
Wynkyn de Worde, 170.

OTHER WORKS BY THE SAME AUTHOR

1. *The Cures of the Diseases in Forraine Attempts of the English Nation*, London, 1598, reproduced in Facsimile with Introduction and Notes. Oxford : Clarendon Press, 1915.

2. *Studies in the History and Method of Science, First Series.* Oxford : Clarendon Press, 1917 (out of print).

3. *Studies in the History and Method of Science, Second Series.* Oxford : Clarendon Press, 1921.

4. *Greek Biology and Greek Medicine.* Oxford : Clarendon Press, 1920.

5. *Greek Science and Modern Science : a Comparison and a Contrast.* London University Press, 1920.

6. *The Discovery of the Circulation of the Blood.* London : G. Bell & Sons, Ltd., 1922.

7. With Professor Henry E. Sigerist, *Essays on the History of Medicine presented to Karl Sudhoff on the occasion of his Seventieth Birthday.* London : Oxford University Press, 1923.

8. With the Hon. Th. Zammit, C.M.G., *Neolithic Representations of the Human Form from the Islands of Malta and Gozo.* The Museum, Valletta, Malta, 1924.

9. With Professor K. Sudhoff. *The Fasciculus Medicinæ of Johannes de Ketham, Alemanus. Facsimile of the First (Venetian) Edition of 1491 with Introduction and Notes.* Large folio. London : Oxford University Press, 1924.

10. With Professor K. Sudhoff. *The Earliest Literature on Syphilis.* Milan : 7 Via Brera, Lier & Co., 1925.

11. *The Fasciculo di Medicina, Venice, 1493, with Introduction, discussion of Art, Language, Sources and Influence, a Translation of the " Anathomia " of Mondino da Luzzi, an account of Mediæval Anatomy and Physiology and an Atlas of Illustrative Figures.* 2 volumes, folio. Milan : 7 Via Brera, Lier & Co., 1925.

CATALOGUE OF DOVER BOOKS

Biological Sciences

AN INTRODUCTION TO GENETICS, A. H. Sturtevant and G. W. Beadle. A very thorough exposition of genetic analysis and the chromosome mechanics of higher organisms by two of the world's most renowned biologists, A. H. Sturtevant, one of the founders of modern genetics, and George Beadle, Nobel laureate in 1958. Does not concentrate on the biochemical approach, but rather more on observed data from experimental evidence and results . . . from Drosophila and other life forms. Some chapter titles: Sex chromosomes; Sex-Linkage; Autosomal Inheritance;; Chromosome Maps; Intra-Chromosomal Rearrangements; Inversions—and Incomplete Chromosomes; Translocations; Lethals; Mutations; Heterogeneous Populations; Genes and Phenotypes; The Determination and Differentiation of Sex; etc. Slightly corrected reprint of 1939 edition. New preface by Drs. Sturtevant and Beadle. 1 color plate. 126 figures. Bibliographies. Index. 391pp. 5⅜ x 8½.   S306 Paperbound **$2.00**

THE GENETICAL THEORY OF NATURAL SELECTION, R. A. Fisher. 2nd revised edition of a vital reviewing of Darwin's Selection Theory in terms of particulate inheritance, by one of the great authorities on experimental and theoretical genetics. Theory is stated in mathematical form. Special features of particulate inheritance are examined: evolution of dominance, maintenance of specific variability, mimicry and sexual selection, etc. 5 chapters on man and his special circumstances as a social animal. 16 photographs. Bibliography. Index. x + 310pp. 5⅜ x 8.   S466 Paperbound **$2.00**

THE ORIENTATION OF ANIMALS: KINESES, TAXES AND COMPASS REACTIONS, Gottfried S. Fraenkel and Donald L. Gunn. A basic work in the field of animal orientations. Complete, detailed survey of everything known in the subject up to 1940s, enlarged and revised to cover major developments to 1960. Analyses of simpler types of orientation are presented in Part I: kinesis, klinotaxis, tropotaxis, telotaxis, etc. Part II covers more complex reactions originating from temperature changes, gravity, chemical stimulation, etc. The two-light experiment and unilateral blinding are dealt with, as is the problem of determinism or volition in lower animals. The book has become the universally-accepted guide to all who deal with the subject—zoologists, biologists, psychologists, and the like. Second, enlarged edition, revised to 1960. Bibliography of over 500 items. 135 illustrations. Indices. xiii + 376pp. 5⅜ x 8½.   T786 Paperbound **$2.25**

THE BEHAVIOUR AND SOCIAL LIFE OF HONEYBEES, C. R. Ribbands. Definitive survey of all aspects of honeybee life and behavior; completely scientific in approach, but written in interesting, everyday language that both professionals and laymen will appreciate. Basic coverage of physiology, anatomy, sensory equipment; thorough account of honeybee behavior in the field (foraging activities, nectar and pollen gathering, how individuals find their way home and back to food areas, mating habits, etc.); details of communication in various field and hive situations. An extensive treatment of activities within the hive community—food sharing, wax production, comb building, swarming, the queen, her life and relationship with the workers, etc. A must for the beekeeper, natural historian, biologist, entomologist, social scientist, et al. "An indispensable reference," J. Hambleton, BĿES. "Recommended in the strongest of terms," AMERICAN SCIENTIST. 9 plates. 66 figures. Indices. 693-item bibliography. 252pp. 5⅜ x 8½.   T1137 Paperbound **$2.00**

BIRD DISPLAY: AN INTRODUCTION TO THE STUDY OF BIRD PSYCHOLOGY, E. A. Armstrong. The standard work on bird display, based on extensive observation by the author and reports of other observers. This important contribution to comparative psychology covers the behavior and ceremonial rituals of hundreds of birds from gannet and heron to birds of paradise and king penguins. Chapters discuss such topics as the ceremonial of the gannet, ceremonial gaping, disablement reactions, the expression of emotions, the evolution and function of social ceremonies, social hierarchy in bird life, dances of birds and men, songs, etc. Free of technical terminology, this work will be equally interesting to psychologists and zoologists as well as bird lovers of all backgrounds. 32 photographic plates. New introduction by the author. List of scientific names of birds. Bibliography. 3-part index. 431pp. 5⅜ x 8½.   T1128 Paperbound **$2.00**

THE SPECIFICITY OF SEROLOGICAL REACTIONS, Karl Landsteiner. With a Chapter on Molecular Structure and Intermolecular Forces by Linus Pauling. Dr. Landsteiner, winner of the Nobel Prize in 1930 for the discovery of the human blood groups, devoted his life to fundamental research and played a leading role in the development of immunology. This authoritative study is an account of the experiments he and his colleagues carried out on antigens and serological reactions with simple compounds. Comprehensive coverage of the basic concepts of immunolgy includes such topics as: The Serological Specificity of Proteins Antigens, Antibodies, Artificially Conjugated Antigens, Non-Protein Cell Substances such as polysaccharides, etc., Antigen-Antibody Reactions (Toxin Neutralization, Precipitin Reactions, Agglutination, etc.). Discussions of toxins, bacterial proteins, viruses, hormones enzymes, etc. in the context of immunological phenomena. New introduction by Dr. Merril Chase of the Rockefeller Institute. Extensive bibliography and bibliography of author's writings. Index. xviii + 330pp. 5⅜ x 8½.   S299 Paperbound **$2.00**

CULTURE METHODS FOR INVERTEBRATE ANIMALS, P. S. Galtsoff, F. E. Lutz, P. S. Welch, J. G. Needham, eds. A compendium of practical experience of hundreds of scientists and technicians, covering invertebrates from protozoa to chordata, in 313 articles on 17 phyla. Explains in great detail food, protection, environment, reproduction conditions, rearing methods, embryology, breeding seasons, schedule of development, much more. Includes at least one species of each considerable group. Half the articles are on class insecta. Introduction. 97 illustrations. Bibliography. Index. xxix + 590pp. 5⅜ x 8. S526 Paperbound **$3.00**

THE BIOLOGY OF THE LABORATORY MOUSE, edited by G. D. Snell. 1st prepared in 1941 by the staff of the Roscoe B. Jackson Memorial Laboratory, this is still the standard treatise on the mouse, assembling an enormous amount of material for which otherwise you spend hours of research. Embryology, reproduction, histology, spontaneous tumor formation, genetics of tumor transplantation, endocrine secretion & tumor formation, milk, influence & tumor formation, inbred, hybrid animals, parasites, infectious diseases, care & recording. Classified bibliography of 1122 items. 172 figures, including 128 photos. ix + 497pp. 6⅛ x 9¼. S248 Clothbound **$6.00**

MATHEMATICAL BIOPHYSICS: PHYSICO-MATHEMATICAL FOUNDATIONS OF BIOLOGY, N. Rashevsky. One of most important books in modern biology, now revised, expanded with new chapters, to include most significant recent contributions. Vol. 1: Diffusion phenomena, particularly diffusion drag forces, their effects. Old theory of cell division based on diffusion drag forces, other theoretical approaches, more exhaustively treated than ever. Theories of excitation, conduction in nerves, with formal theories plus physico-chemical theory. Vol. 2: Mathematical theories of various phenomena in central nervous system. New chapters on theory of color vision, of random nets. Principle of optimal design, extended from earlier edition. Principle of relational mapping of organisms, numerous applications. Introduces into mathematical biology such branches of math as topology, theory of sets. Index. 236 illustrations. Total of 988pp. 5⅜ x 8. S574 Vol. 1 (Books 1, 2) Paperbound **$2.50** / S575 Vol. 2 (Books 3, 4) Paperbound **$2.50** / 2 vol. set **$5.00**

ELEMENTS OF MATHEMATICAL BIOLOGY, A. J. Lotka. A pioneer classic, the first major attempt to apply modern mathematical techniques on a large scale to phenomena of biology, biochemistry, psychology, ecology, similar life sciences. Partial Contents: Statistical meaning of irreversibility; Evolution as redistribution; Equations of kinetics of evolving systems; Chemical, inter-species equilibrium; parameters of state; Energy transformers of nature, etc. Can be read with profit even by those having no advanced math; unsurpassed as study-reference. Formerly titled ELEMENTS OF PHYSICAL BIOLOGY. 72 figures. xxx + 460pp. 5⅜ x 8. S346 Paperbound **$2.45**

THE BIOLOGY OF THE AMPHIBIA, G. K. Noble, Late Curator of Herpetology at the Am. Mus. of Nat. Hist. Probably the most used text on amphibia, unmatched in comprehensiveness, clarity, detail. 19 chapters plus 85-page supplement cover development; heredity; life history; speciation; adaptation; sex, integument, respiratory, circulatory, digestive, muscular, nervous systems; instinct, intelligence, habits, environment, economic value, relationships, classification, etc. "Nothing comparable to it," C. H. Pope, Curator of Amphibia, Chicago Mus. of Nat. Hist. 1047 bibliographic references. 174 illustrations. 600pp. 5⅜ x 8. S206 Paperbound **$2.98**

STUDIES ON THE STRUCTURE AND DEVELOPMENT OF VERTEBRATES, E. S. Goodrich. A definitive study by the greatest modern comparative anatomist. Exceptional in its accounts of the ossicles of the ear, the separate divisions of the coelom and mammalian diaphragm, and the 5 chapters devoted to the head region. Also exhaustive morphological and phylogenetic expositions of skeleton, fins and limbs, skeletal visceral arches and labial cartilages, visceral clefts and gills, vacular, respiratory, excretory, and peripheral nervous systems, etc., from fish to the higher mammals. 754 illustrations. 69 page biographical study by C. C. Hardy. Bibliography of 1186 references. "What an undertaking . . . to write a textbook which will summarize adequately and succinctly all that has been done in the realm of Vertebrate Morphology these recent years," Journal of Anatomy. Index. Two volumes. Total 906pp. 5⅜ x 8. Two vol. set S449-50 Paperbound **$5.00**

A TREATISE ON PHYSIOLOGICAL OPTICS, H. von Helmholtz, Ed. by J. P. C. Southall. Unmatched for thoroughness, soundness, and comprehensiveness, this is still the most important work ever produced in the field of physiological optics. Revised and annotated, it contains everything known about the subject up to 1925. Beginning with a careful anatomical description of the eye, the main body of the text is divided into three general categories: The Dioptrics of the Eye (covering optical imagery, blur circles on the retina, the mechanism of accommodation, chromatic aberration, etc.); The Sensations of Vision (including stimulation of the organ of vision, simple and compound colors, the intensity and duration of light, variations of sensitivity, contrast, etc.); and The Perceptions of Vision (containing movements of the eyes, the monocular field of vision, direction, perception of depth, binocular double vision, etc.). Appendices cover later findings on optical imagery, refraction, ophthalmoscopy, and many other matters. Unabridged, corrected republication of the original English translation of the third German edition. 3 volumes bound as 2. Complete bibliography, 1911-1925. Indices. 312 illustrations. 6 full-page plates, 3 in color. Total of 1,749pp. 5⅜ x 8. Two-volume set S15, 16 Clothbound **$15.00**

INTRODUCTION TO PHYSIOLOGICAL OPTICS, James P. C. Southall, former Professor of Physics in Columbia University. Readable, top-flight introduction, not only for beginning students of optics, but also for other readers—physicists, biochemists, illuminating engineers, optometrists, psychologists, etc. Comprehensive coverage of such matters as the Organ of Vision (structure of the eyeball, the retina, the dioptric system, monocular and binocular vision, adaptation, etc.); The Optical System of the Eye (reflex images in the cornea and crystalline lens, Emmetropia and Ametropia, accommodation, blur circles on retina); Eye-Glasses; Eye Defects; Movements of the Eyeball in its Socket; Rod and Cone Vision; Color Vision; and other similar topics. Index. 134 figures. x +426pp. 5⅜ x 8.   S924 Paperbound **$2.25**

LIGHT, COLOUR AND VISION, Yves LeGrand. A thorough examination of the eye as a receptor of radiant energy and as a mechanism (the retina) consisting of light-sensitive cells which absorb light of various wave lengths—probably the most complete and authoritative treatment of this subject in print. Originally prepared as a series of lectures given at the Institute of Optics in Paris, subsequently enlarged for book publication. Partial contents: Radiant Energy—concept, nature, theories, etc., Sources of Radiation—artificial and natural, the Visual Receptor, Photometric Quantities, Units, Calculations, Retinal Illumination, Trivariance of Vision, Colorimetry, Luminance Difference Thresholds, Anatomy of the Retina, Theories of Vision, Photochemistry and Electro-physiology of the Retina, etc. Appendices, Exercises, with solutions. 500-item bibliography. Authorized translation by R. Hunt, J. Walsh, F. Hunt. Index. 173 illustrations. xiii + 512pp. 5⅜ x 8½.   S979 Clothbound **$10.00**

FINGER PRINTS, PALMS AND SOLES: AN INTRODUCTION TO DERMATOGLYPHICS, Harold Cummins and Charles Midlo. An introduction in non-technical language designed to acquaint the reader with a long-neglected aspect of human biology. Although a chapter dealing with fingerprint identification and the systems of classification used by the FBI, etc. has been added especially for this edition, the main concern of the book is to show how the intricate pattern of ridges and wrinkles on our fingers have a broader significance, applicable in many areas of science and life. Some topics are: the identification of two types of twins; the resolution of doubtful cases of paternity; racial variation; inheritance; the relation of fingerprints to body measurements, blood groups, criminality, character, etc. Classification and recognition of fundamental patterns and pattern types discussed fully. 149 figures. 49 tables. 361-item bibliography. Index. xii + 319pp. 5⅜ x 8⅜.   T778 Paperbound **$2.25**

Classics and histories

ANTONY VAN LEEUWENHOEK AND HIS "LITTLE ANIMALS," edited by Clifford Dobell. First book to treat extensively, accurately, life and works (relating to protozoology, bacteriology) of first microbiologist, bacteriologist, micrologist. Includes founding papers of protozoology, bacteriology; history of Leeuwenhoek's life; discussions of his microscopes, methods, language. His writing conveys sense of an enthusiastic, naive genius, as he looks at rainwater, pepper water, vinegar, frog's skin, rotifers, etc. Extremely readable, even for nonspecialists. "One of the most interesting and enlightening books I have ever read," Dr. C. C. Bass, former Dean, Tulane U. School of Medicine. Only authorized edition. 400-item bibliography. Index. 32 illust. 442pp. 5⅜ x 8.   S594 Paperbound **$2.25**

THE GROWTH OF SCIENTIFIC PHYSIOLOGY, G. J. Goodfield. A compact, superbly written account of how certain scientific investigations brought about the emergence of the distinct science of physiology. Centers principally around the mechanist-vitalist controversy prior to the development of physiology as an independent science, using the arguments which raged around the problem of animal heat as its chief illustration. Covers thoroughly the efforts of clinicians and naturalists and workers in chemistry and physics to solve these problems—from which the new discipline arose. Includes the theories and contributions of: Aristotle, Galen, Harvey, Boyle, Bernard, Benjamin Franklin, Palmer, Gay-Lussac, Priestley, Spallanzani, and many others. 1960 publication. Biographical bibliography. 174pp. 5 x 7½.   T1066 Clothbound **$3.00**

MICROGRAPHIA, Robert Hooke. Hooke, 17th century British universal scientific genius, was a major pioneer in celestial mechanics, optics, gravity, and many other fields, but his greatest contribution was this book, now reprinted entirely from the original 1665 edition, which gave microscopy its first great impetus. With all the freshness of discovery, he describes fully his microscope, and his observations of cork, the edge of a razor, insects' eyes, fabrics, and dozens of other different objects. 38 plates, full-size or larger, contain all the original illustrations. This book is also a fundamental classic in the fields of combustion and heat theory, light and color theory, botany and zoology, hygrometry, and many other fields. It contains such farsighted predictions as the famous anticipation of artificial silk. The final section is concerned with Hooke's telescopic observations of the moon and stars. 323pp. 5⅜ x 8.   T8 Paperbound **$2.50**

Medicine

CLASSICS OF MEDICINE AND SURGERY, edited by C. N. B. Camac. 12 greatest papers in medical history, 11 in full: Lister's "Antiseptic Principle;" Harvey's "Motion in the Heart and Blood;" Auenbrugger's "Percussion of the Chest;" Laënnec's "Auscultation and the Stethoscope;" Jenner's "Inquiry into Smallpox Vaccine," 2 related papers; Morton's "Administering Sulphuric Ether," letters to Warren, "Physiology of Ether;" Simpson's "A New Anaesthetic Agent;" Holmes' "Puerperal Fever." Biographies, portraits of authors, bibliographies. Formerly "Epoch-making Contributions to Medicine, Surgery, and the Allied Sciences." Introduction. 14 illus. 445pp. 5⅜ x 8.   S539 Paperbound **$2.25**

A WAY OF LIFE, Sir William Osler. The complete essay, stating his philosophy of life, as given at Yale University by this great physician and teacher. 30 pages. Copies limited, no more than 1 to a customer. Free.

SOURCE BOOK OF MEDICAL HISTORY, compiled, annotated by Logan Clendening, M.D. Unequalled collection of 139 greatest papers in medical history, by 120 authors, covers almost every area: pathology, asepsis, preventive medicine, bacteriology, physiology, etc. Hippocrates, Gain, Vesalius, Malpighi, Morgagni, Boerhave, Pasteur, Walter Reed, Florence Nightingale, Lavoisier, Claude Bernard, 109 others, give view of medicine unequalled for immediacy. Careful selections give heart of each paper save you reading time. Selections from non-medical literature show lay-views of medicine: Aristophanes, Plato, Arabian Nights, Chaucer, Molière, Dickens, Thackeray, others. "Notable . . . useful to teacher and student alike," Amer. Historical Review. Bibliography. Index. 699pp. 5⅜ x 8.   T621 Paperbound **$2.75**

EXPERIMENTS AND OBSERVATIONS ON THE GASTRIC JUICE AND THE PHYSIOLOGY OF DIGESTION, William Beaumont. A gunshot wound which left a man with a 2½ inch hole through his abdomen into his stomach (1822) enabled Beaumont to perform the remarkable experiments set down here. The first comprehensive, thorough study of motions and processes of the stomach, "his work remains a model of patient, persevering investigation. . . . Beaumont is the pioneer physiologist of this country." (Sir William Osler, in his introduction.) 4 illustrations. xi + 280pp. 5⅜ x 8.   S527 Paperbound **$1.50**

AN INTRODUCTION TO THE STUDY OF EXPERIMENTAL MEDICINE, Claude Bernard. 90-year-old classic of medical science, only major work of Bernard available in English, records his efforts to transform physiology into exact science. Principles of scientific research illustrated by specific case histories from his work; roles of chance, error, preliminary false conclusions, in leading eventually to scientific truth; use of hypothesis. Much of modern application of mathematics to biology rests on the foundation set down here. New foreword by Professor I. B. Cohen, Harvard Univ. xxv + 266pp. 5⅜ x 8.   T400 Paperbound **$1.50**

A WAY OF LIFE, AND OTHER SELECTED WRITINGS, Sir William Osler, Physician and humanist, Osler discourses brilliantly in thought provoking essays and on the history of medicine. He discusses Thomas Browne, Gui Patin, Robert Burton, Michael Servetus, William Beaumont, Laënnec. Includes such favorite writings as the title essay, "The Old Humanities and the New Science," "Creators, Transmitters, and Transmuters," "Books and Men," "The Student Life," and five more of his best discussions of philosophy, religion and literature. 5 photographs. Introduction by G. L. Keynes, M.D., F.R.C.S. Index. xx + 278pp. 5⅜ x 8.
T488 Paperbound **$1.50**

THE HISTORY OF SURGICAL ANESTHESIA, Thomas E. Keys. Concise, but thorough and always engrossing account of the long struggle to find effective methods of eliminating pain during surgery, tracing the remarkable story through the centuries to the eventual successes by dedicated researchers, the acceptance of ether, the work of men such as Priestley, Morton, Lundy, and many, many others. Discussions of the developments in local, regional, and spinal anesthesia, etc. "The general reader as well as the medical historian will find material to interest him in this fascinating story," U.S. QUARTERLY BOOKLIST. Revised, enlarged publication of original edition. Introductory essay by C. D. Leake. Concluding chapter by N. A. Gillespie. Appendix by J. F. Fulton. 46 illustrations. New preface by the author. Chronology of events. Extensive bibliographies. Index. xxx + 193pp. 5⅜ x 8½.
T1122 Paperbound **$2.00**

A SHORT HISTORY OF ANATOMY AND PHYSIOLOGY FROM THE GREEKS TO HARVEY, Charles Singer. Corrected edition of THE EVOLUTION OF ANATOMY, classic work tracing evolution of anatomy and physiology from prescientific times through Greek & Roman periods, Dark Ages, Renaissance, to age of Harvey and beginning of modern concepts. Centered on individuals, movements, periods that definitely advanced anatomical knowledge: Plato, Diocles, Aristotle, Theophrastus, Herophilus, Erasistratus, the Alexandrians, Galen, Mondino, da Vinci, Linacre, Sylvius, others. Special section on Vesalius; Vesalian atlas of nudes, skeletons, muscle tabulae. Index of names, 20 plates. 270 extremely interesting illustrations of ancient, medieval, Renaissance, Oriental origin. xii + 209pp. 5⅜ x 8.   T389 Paperbound **$1.75**

Books Explaining Science and Mathematics

WHAT IS SCIENCE?, N. Campbell. The role of experiment and measurement, the function of mathematics, the nature of scientific laws, the difference between laws and theories, the limitations of science, and many similarly provocative topics are treated clearly and without technicalities by an eminent scientist. "Still an excellent introduction to scientific philosophy," H. Margenau in PHYSICS TODAY. "A first-rate primer . . . deserves a wide audience," SCIENTIFIC AMERICAN. 192pp. 5⅜ x 8.　　S43 Paperbound **$1.25**

THE NATURE OF PHYSICAL THEORY, P. W. Bridgman. A Nobel Laureate's clear, non-technical lectures on difficulties and paradoxes connected with frontier research on the physical sciences. Concerned with such central concepts as thought, logic, mathematics, relativity, probability, wave mechanics, etc. he analyzes the contributions of such men as Newton, Einstein, Bohr, Heisenberg, and many others. "Lucid and entertaining . . . recommended to anyone who wants to get some insight into current philosophies of science," THE NEW PHILOSOPHY. Index. xi + 138pp. 5⅜ x 8.　　S33 Paperbound **$1.25**

EXPERIMENT AND THEORY IN PHYSICS, Max Born. A Nobel Laureate examines the nature of experiment and theory in theoretical physics and analyzes the advances made by the great physicists of our day: Heisenberg, Einstein, Bohr, Planck, Dirac, and others. The actual process of creation is detailed step-by-step by one who participated. A fine examination of the scientific method at work. 44pp. 5⅜ x 8.　　S308 Paperbound **75¢**

THE PSYCHOLOGY OF INVENTION IN THE MATHEMATICAL FIELD, J. Hadamard. The reports of such men as Descartes, Pascal, Einstein, Poincaré, and others are considered in this investigation of the method of idea-creation in mathematics and other sciences and the thinking process in general. How do ideas originate? What is the role of the unconscious? What is Poincaré's forgetting hypothesis? are some of the fascinating questions treated. A penetrating analysis of Einstein's thought processes concludes the book. xiii + 145pp. 5⅜ x 8.　　T107 Paperbound **$1.25**

THE NATURE OF LIGHT AND COLOUR IN THE OPEN AIR, M. Minnaert. Why are shadows sometimes blue, sometimes green, or other colors depending on the light and surroundings? What causes mirages? Why do multiple suns and moons appear in the sky? Professor Minnaert explains these unusual phenomena and hundreds of others in simple, easy-to-understand terms based on optical laws and the properties of light and color. No mathematics is required but artists, scientists, students, and everyone fascinated by these "tricks" of nature will find thousands of useful and amazing pieces of information. Hundreds of observational experiments are suggested which require no special equipment. 200 illustrations; 42 photos. xvi + 362pp. 5⅜ x 8.　　T196 Paperbound **$2.00**

*MATHEMATICS IN ACTION, O. G. Sutton. Everyone with a command of high school algebra will find this book one of the finest possible introductions to the application of mathematics to physical theory. Ballistics, numerical analysis, waves and wavelike phenomena, Fourier series, group concepts, fluid flow and aerodynamics, statistical measures, and meteorology are discussed with unusual clarity. Some calculus and differential equations theory is developed by the author for the reader's help in the more difficult sections. 88 figures. Index. viii + 236pp. 5⅜ x 8.　　T440 Clothbound **$3.50**

SOAP-BUBBLES: THEIR COLOURS AND THE FORCES THAT MOULD THEM, C. V. Boys. For continuing popularity and validity as scientific primer, few books can match this volume of easily-followed experiments, explanations. Lucid exposition of complexities of liquid films, surface tension and related phenomena, bubbles' reaction to heat, motion, music, magnetic fields. Experiments with capillary attraction, soap bubbles on frames, composite bubbles, liquid cylinders and jets, bubbles other than soap, etc. Wonderful introduction to scientific method, natural laws that have many ramifications in areas of modern physics. Only complete edition in print. New Introduction by S. Z. Lewin, New York University. 83 illustrations; 1 full-page color plate. xii + 190pp. 5⅜ x 8½.　　T542 Paperbound **95¢**

CATALOGUE OF DOVER BOOKS

*THE EVOLUTION OF SCIENTIFIC THOUGHT FROM NEWTON TO EINSTEIN, A. d'Abro. A detailed account of the evolution of classical physics into modern relativistic theory and the concommitant changes in scientific methodology. The breakdown of classical physics in the face of non-Euclidean geometry and the electromagnetic equations is carefully discussed and then an exhaustive analysis of Einstein's special and general theories of relativity and their implications is given. Newton, Riemann, Weyl, Lorentz, Planck, Maxwell, and many others are considered. A non-technical explanation of space, time, electromagnetic waves, etc. as understood today. "Model of semi-popular exposition," NEW REPUBLIC. 21 diagrams. 482pp. 5⅜ x 8.
T2 Paperbound **$2.25**

EINSTEIN'S THEORY OF RELATIVITY, Max Born. Nobel Laureate explains Einstein's special and general theories of relativity, beginning with a thorough review of classical physics in simple, non-technical language. Discussion of Einstein's work discusses concept of simultaneity, kinematics, relativity of arbitrary motions, the space-time continuum, geometry of curved surfaces, etc., steering middle course between vague popularizations and complex scientific presentations. 1962 edition revised by author takes into account latest findings, predictions of theory and implications for cosmology, indicates what is being sought in unified field theory. Mathematics very elementary, illustrative diagrams and experiments informative but simple. Revised 1962 edition. Revised by Max Born, assisted by Gunther Leibfried and Walter Biem. Index. 143 illustrations. vii + 376pp. 5⅜ x 8.
S769 Paperbound **$2.00**

PHILOSOPHY AND THE PHYSICISTS, L. Susan Stebbing. A philosopher examines the philosophical aspects of modern science, in terms of a lively critical attack on the ideas of Jeans and Eddington. Such basic questions are treated as the task of science, causality, determinism, probability, consciousness, the relation of the world of physics to the world of everyday experience. The author probes the concepts of man's smallness before an inscrutable universe, the tendency to idealize mathematical construction, unpredictability theorems and human freedom, the supposed opposition between 19th century determinism and modern science, and many others. Introduces many thought-stimulating ideas about the implications of modern physical concepts. xvi + 295pp. 5⅜ x 8.
T480 Paperbound **$1.65**

THE RESTLESS UNIVERSE, Max Born. A remarkably lucid account by a Nobel Laureate of recent theories of wave mechanics, behavior of gases, electrons and ions, waves and particles, electronic structure of the atom, nuclear physics, and similar topics. "Much more thorough and deeper than most attempts . . . easy and delightful," CHEMICAL AND ENGINEERING NEWS. Special feature: 7 animated sequences of 60 figures each showing such phenomena as gas molecules in motion, the scattering of alpha particles, etc. 11 full-page plates of photographs. Total of nearly 600 illustrations. 351pp. 6⅛ x 9¼.
T412 Paperbound **$2.00**

THE COMMON SENSE OF THE EXACT SCIENCES, W. K. Clifford. For 70 years a guide to the basic concepts of scientific and mathematical thought. Acclaimed by scientists and laymen alike, it offers a wonderful insight into concepts such as the extension of meaning of symbols, characteristics of surface boundaries, properties of plane figures, measurement of quantities, vectors, the nature of position, bending of space, motion, mass and force, and many others. Prefaces by Bertrand Russell and Karl Pearson. Critical introduction by James Newman. 130 figures. 249pp. 5⅜ x 8.
T61 Paperbound **$1.60**

MATTER AND LIGHT, THE NEW PHYSICS, Louis de Broglie. Non-technical explanations by a Nobel Laureate of electro-magnetic theory, relativity, matter, light and radiation, wave mechanics, quantum physics, philosophy of science, and similar topics. This is one of the simplest yet most accurate introductions to the work of men like Planck, Einstein, Bohr, and others. Only 2 of the 21 chapters require a knowledge of mathematics. 300pp. 5⅜ x 8.
T35 Paperbound **$1.85**

SCIENCE, THEORY AND MAN, Erwin Schrödinger. This is a complete and unabridged reissue of SCIENCE AND THE HUMAN TEMPERAMENT plus an additional essay: "What Is an Elementary Particle?" Nobel Laureate Schrödinger discusses such topics as nature of scientific method, the nature of science, chance and determinism, science and society, conceptual models for physical entities, elementary particles and wave mechanics. Presentation is popular and may be followed by most people with little or no scientific training. "Fine practical preparation for a time when laws of nature, human institutions . . . are undergoing a critical examination without parallel," Waldemar Kaempffert, N. Y. TIMES. 192pp. 5⅜ x 8.
T428 Paperbound **$1.35**

CONCERNING THE NATURE OF THINGS, Sir William Bragg. The Nobel Laureate physicist in his Royal Institute Christmas Lectures explains such diverse phenomena as the formation of crystals, how uranium is transmuted to lead, the way X-rays work, why a spinning ball travels in a curved path, the reason why bubbles bounce from each other, and many other scientific topics that are seldom explained in simple terms. No scientific background needed—book is easy enough that any intelligent adult or youngster can understand it. Unabridged. 32pp. of photos; 57 figures. xii + 232pp. 5⅜ x 8.
T31 Paperbound **$1.35**

*THE RISE OF THE NEW PHYSICS (formerly THE DECLINE OF MECHANISM), A. d'Abro. This authoritative and comprehensive 2 volume exposition is unique in scientific publishing. Written for intelligent readers not familiar with higher mathematics, it is the only thorough explanation in non-technical language of modern mathematical-physical theory. Combining both history and exposition, it ranges from classical Newtonian concepts up through the electronic theories of Dirac and Heisenberg, the statistical mechanics of Fermi, and Einstein's relativity theories. "A must for anyone doing serious study in the physical sciences," J. OF FRANKLIN INST. 97 illustrations. 991pp. 2 volumes.
T3 Vol. 1, Paperbound **$2.25**
T4 Vol. 2, Paperbound **$2.25**

SCIENCE AND HYPOTHESIS, Henri Poincaré. Creative psychology in science. How such concepts as number, magnitude, space, force, classical mechanics were developed and how the modern scientist uses them in his thought. Hypothesis in physics, theories of modern physics. Introduction by Sir James Larmor. "Few mathematicians have had the breadth of vision of Poincaré, and none is his superior in the gift of clear exposition," E. T. Bell. Index. 272pp. 5⅜ x 8.
S221 Paperbound **$1.35**

THE VALUE OF SCIENCE, Henri Poincaré. Many of the most mature ideas of the "last scientific universalist" conveyed with charm and vigor for both the beginning student and the advanced worker. Discusses the nature of scientific truth, whether order is innate in the universe or imposed upon it by man, logical thought versus intuition (relating to mathematics through the works of Weierstrass, Lie, Klein, Riemann), time and space (relativity, psychological time, simultaneity), Hertz's concept of force, interrelationship of mathematical physics to pure math, values within disciplines of Maxwell, Carnot, Mayer, Newton, Lorentz, etc. Index. iii + 147pp. 5⅜ x 8.
S469 Paperbound **$1.35**

THE SKY AND ITS MYSTERIES, E. A. Beet. One of the most lucid books on the mysteries of the universe; covers history of astronomy from earliest observations to modern theories of expanding universe, source of stellar energy, birth of planets, origin of moon craters, possibilities of life on other planets. Discusses effects of sunspots on weather; distance, age of stars; methods and tools of astronomers; much more. Expert and fascinating. "Eminently readable book," London Times. Bibliography. Over 50 diagrams, 12 full-page plates. Fold-out star map. Introduction. Index. 238pp. 5¼ x 7½.
T627 Clothbound **$3.50**

OUT OF THE SKY: AN INTRODUCTION TO METEORITICS, H. H. Nininger. A non-technical yet comprehensive introduction to the young science of meteoritics: all aspects of the arrival of cosmic matter on our planet from outer space and the reaction and alteration of this matter in the terrestrial environment. Essential facts and major theories presented by one of the world's leading experts. Covers ancient reports of meteors; modern systematic investigations; fireball clusters; meteorite showers; tektites; planetoidal encounters; etc. 52 full-page plates with over 175 photographs. 22 figures. Bibliography and references. Index. viii + 336pp. 5⅜ x 8.
T519 Paperbound **$1.85**

THE REALM OF THE NEBULAE, E. Hubble. One of great astronomers of our day records his formulation of concept of "island universes." Covers velocity-distance relationship; classification, nature, distances, general types of nebulae; cosmological theories. A fine introduction to modern theories for layman. No math needed. New introduction by A. Sandage. 55 illustrations, photos. Index. iv + 201pp. 5⅜ x 8.
S455 Paperbound **$1.50**

AN ELEMENTARY SURVEY OF CELESTIAL MECHANICS, Y. Ryabov. Elementary exposition of gravitational theory and celestial mechanics. Historical introduction and coverage of basic principles, including: the ecliptic, the orbital plane, the 2- and 3-body problems, the discovery of Neptune, planetary rotation, the length of the day, the shapes of galaxies, satellites (detailed treatment of Sputnik I), etc. First American reprinting of successful Russian popular exposition. Follow actual methods of astrophysicists with only high school math! Appendix. 58 figures. 165pp. 5⅜ x 8.
T756 Paperbound **$1.25**

GREAT IDEAS AND THEORIES OF MODERN COSMOLOGY, Jagjit Singh. Companion volume to author's popular "Great Ideas of Modern Mathematics" (Dover, $1.55). The best non-technical survey of post-Einstein attempts to answer perhaps unanswerable questions of origin, age of Universe, possibility of life on other worlds, etc. Fundamental theories of cosmology and cosmogony recounted, explained, evaluated in light of most recent data: Einstein's concepts of relativity, space-time; Milne's a priori world-system; astrophysical theories of Jeans, Eddington; Hoyle's "continuous creation;" contributions of dozens more scientists. A faithful, comprehensive critical summary of complex material presented in an extremely well-written text intended for laymen. Original publication. Index. xii + 276pp. 5⅜ x 8½.
T925 Paperbound **$1.85**

BASIC ELECTRICITY, Bureau of Naval Personnel. Very thorough, easily followed course in basic electricity for beginner, layman, or intermediate student. Begins with simplest definitions, presents coordinated, systematic coverage of basic theory and application: conductors, insulators, static electricity, magnetism, production of voltage, Ohm's law, direct current series and parallel circuits, wiring techniques, electromagnetism, alternating current, capacitance and inductance, measuring instruments, etc.; application to electrical machines such as alternating and direct current generators, motors, transformers, magnetic magnifiers, etc. Each chapter contains problems to test progress; answers at rear. No math needed beyond algebra. Appendices on signs, formulas, etc. 345 illustrations. 448pp. 7½ x 10.
S973 Paperbound **$3.00**

ELEMENTARY METALLURGY AND METALLOGRAPHY, A. M. Shrager. An introduction to common metals and alloys; stress is upon steel and iron, but other metals and alloys also covered. All aspects of production, processing, working of metals. Designed for student who wishes to enter metallurgy, for bright high school or college beginner, layman who wants background on extremely important industry. Questions, at ends of chapters, many microphotographs, glossary. Greatly revised 1961 edition. 195 illustrations, tables. ix + 389pp. 5⅜ x 8.
S138 Paperbound **$2.25**

BRIDGES AND THEIR BUILDERS, D. B. Steinman & S. R. Watson. Engineers, historians, and every person who has ever been fascinated by great spans will find this book an endless source of information and interest. Greek and Roman structures, Medieval bridges, modern classics such as the Brooklyn Bridge, and the latest developments in the science are retold by one of the world's leading authorities on bridge design and construction. BRIDGES AND THEIR BUILDERS is the only comprehensive and accurate semi-popular history of tnese important measures of progress in print. New, greatly revised, enlarged edition. 23 photos; 26 line-drawings. Index. xvii + 401pp. 5⅜ x 8.   T431 Paperbound **$2.00**

FAMOUS BRIDGES OF THE WORLD, D. B. Steinman. An up-to-the-minute new edition of a book that explains the fascinating drama of how the world's great bridges came to be built. The author, designer of the famed Mackinac bridge, discusses bridges from all periods and all parts of the world, explaining their various types of construction, and describing the problems their builders faced. Although primarily for youngsters, this cannot fail to interest readers of all ages. 48 illustrations in the text. 23 photographs. 99pp. 6⅛ x 9¼.   T161 Paperbound **$1.00**

HOW DO YOU USE A SLIDE RULE? by A. A. Merrill. A step-by-step explanation of the slide rule that presents the fundamental rules clearly enough for the non-mathematician to understand. Unlike most instruction manuals, this work concentrates on the two most important operations: multiplication and division. 10 easy lessons, each with a clear drawing, for the reader who has difficulty following other expositions. 1st publication. Index. 2 Appendices. 10 illustrations. 78 problems, all with answers. vi + 36 pp. 6⅛ x 9¼.   T62 Paperbound **60¢**

HOW TO CALCULATE QUICKLY, H. Sticker. A tried and true method for increasing your "number sense" — the ability to see relationships between numbers and groups of numbers. Addition, subtraction, multiplication, division, fractions, and other topics are treated through techniques not generally taught in schools: left to right multiplication, division by inspection, etc. This is not a collection of tricks which work only on special numbers, but a detailed well-planned course, consisting of over 9,000 problems that you can work in spare moments. It is excellent for anyone who is inconvenienced by slow computational skills. 5 or 10 minutes of this book daily will double or triple your calculation speed. 9,000 problems, answers. 256pp. 5⅜ x 8.   T295 Paperbound **$1.00**

MATHEMATICAL FUN, GAMES AND PUZZLES, Jack Frohlichstein. A valuable service for parents of children who have trouble with math, for teachers in need of a supplement to regular upper elementary and junior high math texts (each section is graded—easy, average, difficult —for ready adaptation to different levels of ability), and for just anyone who would like to develop basic skills in an informal and entertaining manner. The author combines ten years of experience as a junior high school math teacher with a method that uses puzzles and games to introduce the basic ideas and operations of arithmetic. Stress on everyday uses of math: banking, stock market, personal budgets, insurance, taxes. Intellectually stimulating and practical, too. 418 problems and diversions with answers. Bibliography. 120 illustrations. xix + 306pp. 5⅝ x 8½.   T789 Paperbound **$1.75**

GREAT IDEAS OF MODERN MATHEMATICS: THEIR NATURE AND USE, Jagjit Singh. Reader with only high school math will understand main mathematical ideas of modern physics, astronomy, genetics, psychology, evolution, etc. better than many who use them as tools, but comprehend little of their basic structure. Author uses his wide knowledge of non-mathematical fields in brilliant exposition of differential equations, matrices, group theory, logic, statistics, problems of mathematical foundations, imaginary numbers, vectors, etc. Original publication. 2 appendixes. 2 indexes. 65 illustr. 322pp. 5⅜ x 8.   S587 Paperbound **$1.75**

THE UNIVERSE OF LIGHT, W. Bragg. Sir William Bragg, Nobel Laureate and great modern physicist, is also well known for his powers of clear exposition. Here he analyzes all aspects of light for the layman: lenses, reflection, refraction, the optics of vision, x-rays, the photoelectric effect, etc. He tells you what causes the color of spectra, rainbows, and soap bubbles, how magic mirrors work, and much more. Dozens of simple experiments are described. Preface. Index. 199 line drawings and photographs, including 2 full-page color plates. x + 283pp. 5⅜ x 8.   T538 Paperbound **$1.85**

*****INTRODUCTION TO SYMBOLIC LOGIC AND ITS APPLICATIONS, Rudolph Carnap.** One of the clearest, most comprehensive, and rigorous introductions to modern symbolic logic, by perhaps its greatest living master. Not merely elementary theory, but demonstrated applications in mathematics, physics, and biology. Symbolic languages of various degrees of complexity are analyzed, and one constructed. "A creation of the rank of a masterpiece," Zentralblatt für Mathematik und Ihre Grenzgebiete. Over 300 exercises. 5 figures. Bibliography. Index. xvi + 241pp. 5⅜ x 8.   S453 Paperbound **$1.85**

*****HIGHER MATHEMATICS FOR STUDENTS OF CHEMISTRY AND PHYSICS, J. W. Mellor.** Not abstract, but practical, drawing its problems from familiar laboratory material, this book covers theory and application of differential calculus, analytic geometry, functions with singularities, integral calculus, infinite series, solution of numerical equations, differential equations, Fourier's theorem and extensions, probability and the theory of errors, calculus of variations, determinants, etc. "If the reader is not familiar with this book, it will repay him to examine it," CHEM. & ENGINEERING NEWS. 800 problems. 189 figures. 2 appendices; 30 tables of integrals, probability functions, etc. Bibliography. xxi + 641pp. 5⅜ x 8.
S193 Paperbound **$2.50**

CATALOGUE OF DOVER BOOKS

THE FOURTH DIMENSION SIMPLY EXPLAINED, edited by Henry P. Manning. Originally written as entries in contest sponsored by "Scientific American," then published in book form, these 22 essays present easily understood explanations of how the fourth dimension may be studied, the relationship of non-Euclidean geometry to the fourth dimension, analogies to three-dimensional space, some fourth-dimensional absurdities and curiosities, possible measurements and forms in the fourth dimension. In general, a thorough coverage of many of the simpler properties of fourth-dimensional space. Multi-points of view on many of the most important aspects are valuable aid to comprehension. Introduction by Dr. Henry P. Manning gives proper emphasis to points in essays, more advanced account of fourth-dimensional geometry. 82 figures. 251pp. 5⅜ x 8. T711 Paperbound **$1.35**

TRIGONOMETRY REFRESHER FOR TECHNICAL MEN, A. A. Klaf. A modern question and answer text on plane and spherical trigonometry. Part I covers plane trigonometry: angles, quadrants, trigonometrical functions, graphical representation, interpolation, equations, logarithms, solution of triangles, slide rules, etc. Part II discusses applications to navigation, surveying, elasticity, architecture, and engineering. Small angles, periodic functions, vectors, polar coordinates, De Moivre's theorem, fully covered. Part III is devoted to spherical trigonometry and the solution of spherical triangles, with applications to terrestrial and astronomical problems. Special time-savers for numerical calculation. 913 questions answered for you! 1738 problems; answers to odd numbers. 494 figures. 14 pages of functions, formulae. Index. x + 629pp. 5⅜ x 8. T371 Paperbound **$2.00**

CALCULUS REFRESHER FOR TECHNICAL MEN. A. A. Klaf. Not an ordinary textbook but a unique refresher for engineers, technicians, and students. An examination of the most important aspects of differential and integral calculus by means of 756 key questions. Part I covers simple differential calculus: constants, variables, functions, increments, derivatives, logarithms, curvature, etc. Part II treats fundamental concepts of integration: inspection, substitution, transformation, reduction, areas and volumes, mean value, successive and partial integration, double and triple integration. Stresses practical aspects! A 50 page section gives applications to civil and nautical engineering, electricity, stress and strain, elasticity, industrial engineering, and similar fields. 756 questions answered. 556 problems; solutions to odd numbers. 36 pages of constants, formulae. Index. v + 431pp. 5⅜ x 8. T370 Paperbound **$2.00**

PROBABILITIES AND LIFE, Emile Borel. One of the leading French mathematicians of the last 100 years makes use of certain results of mathematics of probabilities and explains a number of problems that for the most part, are related to everyday living or to illness and death: computation of life expectancy tables, chances of recovery from various diseases, probabilities of job accidents, weather predictions, games of chance, and so on. Emphasis on results not processes, though some indication is made of mathematical proofs. Simple in style, free of technical terminology, limited in scope to everyday situations, it is comprehensible to laymen, fine reading for beginning students of probability. New English translation. Index. Appendix. vi + 87pp. 5⅜ x 8½. T121 Paperbound **$1.00**

POPULAR SCIENTIFIC LECTURES, Hermann von Helmholtz. 7 lucid expositions by a pre-eminent scientific mind: "The Physiological Causes of Harmony in Music," "On the Relation of Optics to Painting," "On the Conservation of Force," "On the Interaction of Natural Forces," "On Goethe's Scientific Researches" into theory of color, "On the Origin and Significance of Geometric Axioms," "On Recent Progress in the Theory of Vision." Written with simplicity of expression, stripped of technicalities, these are easy to understand and delightful reading for anyone interested in science or looking for an introduction to serious study of acoustics or optics. Introduction by Professor Morris Kline, Director, Division of Electromagnetic Research, New York University, contains astute, impartial evaluations. Selected from "Popular Lectures on Scientific Subjects," 1st and 2nd series. xii + 286pp. 5⅜ x 8½. T799 Paperbound **$1.45**

SCIENCE AND METHOD, Henri Poincaré. Procedure of scientific discovery, methodology, experiment, idea-germination—the intellectual processes by which discoveries come into being. Most significant and most interesting aspects of development, application of ideas. Chapters cover selection of facts, chance, mathematical reasoning, mathematics, and logic; Whitehead, Russell, Cantor; the new mechanics, etc. 288pp. 5⅜ x 8. S222 Paperbound **$1.50**

HEAT AND ITS WORKINGS, Morton Mott-Smith, Ph.D. An unusual book; to our knowledge the only middle-level survey of this important area of science. Explains clearly such important concepts as physiological sensation of heat and Weber's law, measurement of heat, evolution of thermometer, nature of heat, expansion and contraction of solids, Boyle's law, specific heat. BTU's and calories, evaporation, Andrews's isothermals, radiation, the relation of heat to light, many more topics inseparable from other aspects of physics. A wide, non-mathematical yet thorough explanation of basic ideas, theories, phenomena for laymen and beginning scientists illustrated by experiences of daily life. Bibliography. 50 illustrations. x + 165pp. 5⅜ x 8½. T978 Paperbound **$1.00**

History of Science and Mathematics

THE STUDY OF THE HISTORY OF MATHEMATICS, THE STUDY OF THE HISTORY OF SCIENCE, G. Sarton. Two books bound as one. Each volume contains a long introduction to the methods and philosophy of each of these historical fields, covering the skills and sympathies of the historian, concepts of history of science, psychology of idea-creation, and the purpose of history of science. Prof. Sarton also provides more than 80 pages of classified bibliography. Complete and unabridged. Indexed. 10 illustrations. 188pp. 5⅜ x 8.   T240 Paperbound **$1.25**

A HISTORY OF PHYSICS, Florian Cajori, Ph.D. First written in 1899, thoroughly revised in 1929, this is still best entry into antecedents of modern theories. Precise non-mathematical discussion of ideas, theories, techniques, apparatus of each period from Greeks to 1920's, analyzing within each period basic topics of matter, mechanics, light, electricity and magnetism, sound, atomic theory, etc. Stress on modern developments, from early 19th century to present. Written with critical eye on historical development, significance. Provides most of needed historical background for student of physics. Reprint of second (1929) edition. Index. Bibliography in footnotes. 16 figures. xv + 424pp. 5⅜ x 8.   T970 Paperbound **$2.00**

A HISTORY OF ASTRONOMY FROM THALES TO KEPLER, J. L. E. Dreyer. Formerly titled A HISTORY OF PLANETARY SYSTEMS FROM THALES TO KEPLER. This is the only work in English which provides a detailed history of man's cosmological views from prehistoric times up through the Renaissance. It covers Egypt, Babylonia, early Greece, Alexandria, the Middle Ages, Copernicus, Tycho Brahe, Kepler, and many others. Epicycles and other complex theories of positional astronomy are explained in terms nearly everyone will find clear and easy to understand. "Standard reference on Greek astronomy and the Copernican revolution," SKY AND TELESCOPE. Bibliography. 21 diagrams. Index. xvii + 430pp. 5⅜ x 8.   S79 Paperbound **$2.25**

A SHORT HISTORY OF ASTRONOMY, A. Berry. A popular standard work for over 50 years, this thorough and accurate volume covers the science from primitive times to the end of the 19th century. After the Greeks and Middle Ages, individual chapters analyze Copernicus, Brahe, Galileo, Kepler, and Newton, and the mixed reception of their startling discoveries. Post-Newtonian achievements are then discussed in unusual detail: Halley, Bradley, Lagrange, Laplace, Herschel, Bessel, etc. 2 indexes. 104 illustrations, 9 portraits. xxxi + 440pp. 5⅜ x 8.   T210 Paperbound **$2.00**

PIONEERS OF SCIENCE, Sir Oliver Lodge. An authoritative, yet elementary history of science by a leading scientist and expositor. Concentrating on individuals—Copernicus, Brahe, Kepler, Galileo, Descartes, Newton, Laplace, Herschel, Lord Kelvin, and other scientists—the author presents their discoveries in historical order, adding biographical material on each man and full, specific explanations of their achievements. The full, clear discussions of the accomplishments of post-Newtonian astronomers are features seldom found in other books on the subject. Index. 120 illustrations. xv + 404pp. 5⅜ x 8.   T716 Paperbound **$1.65**

THE BIRTH AND DEVELOPMENT OF THE GEOLOGICAL SCIENCES, F. D. Adams. The most complete and thorough history of the earth sciences in print. Geological thought from earliest recorded times to the end of the 19th century—covers over 300 early thinkers and systems: fossils and hypothetical explanations of them, vulcanists vs. neptunists, figured stones and paleontology, generation of stones, and similar topics. 91 illustrations, including medieval, renaissance woodcuts, etc. 632 footnotes and bibliographic notes. Index. 511pp. 5⅜ x 8.
T5 Paperbound **$2.25**

THE STORY OF ALCHEMY AND EARLY CHEMISTRY, J. M. Stillman. "Add the blood of a red-haired man"—a recipe typical of the many quoted in this authoritative and readable history of the strange beliefs and practices of the alchemists. Concise studies of every leading figure in alchemy and early chemistry through Lavoisier, in this curious epic of superstition and true science, constructed from scores of rare and difficult Greek, Latin, German, and French texts. Foreword by S. W. Young. 246-item bibliography. Index. xiii + 566pp. 5⅜ x 8.
S628 Paperbound **$2.45**

HISTORY OF MATHEMATICS, D. E. Smith. Most comprehensive non-technical history of math in English. Discusses the lives and works of over a thousand major and minor figures, from Euclid to Descartes, Gauss, and Riemann. Vol. I: A chronological examination, from primitive concepts through Egypt, Babylonia, Greece, the Orient, Rome, the Middle Ages, the Renaissance, and up to 1900. Vol. 2: The development of ideas in specific fields and problems, up through elementary calculus. Two volumes, total of 510 illustrations, 1355pp. 5⅜ x 8. Set boxed in attractive container.   T429,430 Paperbound the set **$5.00**

Classics of Science

THE DIDEROT PICTORIAL ENCYCLOPEDIA OF TRADES AND INDUSTRY, MANUFACTURING AND THE TECHNICAL ARTS IN PLATES SELECTED FROM "L'ENCYCLOPEDIE OU DICTIONNAIRE RAISONNE DES SCIENCES, DES ARTS, ET DES METIERS" OF DENIS DIDEROT, edited with text by C. Gillispie. The first modern selection of plates from the high point of 18th century French engraving, Diderot's famous Encyclopedia. Over 2000 illustrations on 485 full page plates, most of them original size, illustrating the trades and industries of one of the most fascinating periods of modern history, 18th century France. These magnificent engravings provide an invaluable glimpse into the past for the student of early technology, a lively and accurate social document to students of cultures, an outstanding find to the lover of fine engravings. The plates teem with life, with men, women, and children performing all of the thousands of operations necessary to the trades before and during the early stages of the industrial revolution. Plates are in sequence, and show general operations, closeups of difficult operations, and details of complex machinery. Such important and interesting trades and industries are illustrated as sowing, harvesting, beekeeping, cheesemaking, operating windmills, milling flour, charcoal burning, tobacco processing, indigo, fishing, arts of war, salt extraction, mining, smelting iron, casting iron, steel, extracting mercury, zinc, sulphur, copper, etc., slating, tinning, silverplating, gilding, making gunpowder, cannons, bells, shoeing horses, tanning, papermaking, printing, dying, and more than 40 other categories. 920pp. 9 x 12. Heavy library cloth. T421 Two volume set **$18.50**

THE PRINCIPLES OF SCIENCE, A TREATISE ON LOGIC AND THE SCIENTIFIC METHOD, W. Stanley Jevons. Treating such topics as Inductive and Deductive Logic, the Theory of Number, Probability, and the Limits of Scientific Method, this milestone in the development of symbolic logic remains a stimulating contribution to the investigation of inferential validity in the natural and social sciences. It significantly advances Boole's logic, and describes a machine which is a foundation of modern electronic calculators. In his introduction, Ernest Nagel of Columbia University says, "(Jevons) . . . continues to be of interest as an attempt to articulate the logic of scientific inquiry." Index. liii + 786pp. 5⅜ x 8. S446 Paperbound **$2.98**

*DIALOGUES CONCERNING TWO NEW SCIENCES, Galileo Galilei. A classic of experimental science which has had a profound and enduring influence on the entire history of mechanics and engineering. Galileo based this, his finest work, on 30 years of experimentation. It offers a fascinating and vivid exposition of dynamics, elasticity, sound, ballistics, strength of materials, and the scientific method. Translated by H. Crew and A. de Salvio. 126 diagrams. Index. xxi + 288pp. 5⅜ x 8. S99 Paperbound **$1.75**

DE MAGNETE, William Gilbert. This classic work on magnetism founded a new science. Gilbert was the first to use the word "electricity," to recognize mass as distinct from weight, to discover the effect of heat on magnetic bodies; invented an electroscope, differentiated between static electricity and magnetism, conceived of the earth as a magnet. Written by the first great experimental scientist, this lively work is valuable not only as an historical landmark, but as the delightfully easy-to-follow record of a perpetually searching, ingenious mind. Translated by P. F. Mottelay. 25 page biographical memoir. 90 fix. lix + 368pp. 5⅜ x 8. S470 Paperbound **$2.00**

*OPTICKS, Sir Isaac Newton. An enormous storehouse of insights and discoveries on light, reflection, color, refraction, theories of wave and corpuscular propagation of light, optical apparatus, and mathematical devices which have recently been reevaluated in terms of modern physics and placed in the top-most ranks of Newton's work! Foreword by Albert Einstein. Preface by I. B. Cohen of Harvard U. 7 pages of portraits, facsimile pages, letters, etc. cxvi + 412pp. 5⅜ x 8. S205 Paperbound **$2.25**

A SURVEY OF PHYSICAL THEORY, M. Planck. Lucid essays on modern physics for the general reader by the Nobel Laureate and creator of the quantum revolution. Planck explains how the new concepts came into being; explores the clash between theories of mechanics, electrodynamics, and thermodynamics; and traces the evolution of the concept of light through Newton, Huygens, Maxwell, and his own quantum theory, providing unparalleled insights into his development of this momentous modern concept. Bibliography. Index. vii + 121pp. 5⅜ x 8. S650 Paperbound **$1.15**

A SOURCE BOOK IN MATHEMATICS, D. E. Smith. English translations of the original papers that announced the great discoveries in mathematics from the Renaissance to the end of the 19th century: succinct selections from 125 different treatises and articles, most of them unavailable elsewhere in English—Newton, Leibniz, Pascal, Riemann, Bernoulli, etc. 24 articles trace developments in the field of number, 18 cover algebra, 36 are on geometry, and 13 on calculus. Biographical-historical introductions to each article. Two volume set. Index in each. Total of 115 illustrations. Total of xxviii + 742pp. 5⅜ x 8. S552 Vol I Paperbound **$2.00**
S553 Vol II Paperbound **$2.00**
The set, boxed **$4.00**

*THE THIRTEEN BOOKS OF EUCLID'S ELEMENTS, edited by T. L. Heath. This is the complete EUCLID — the definitive edition of one of the greatest classics of the western world. Complete English translation of the Heiberg text with spurious Book XIV. Detailed 150-page introduction discusses aspects of Greek and medieval mathematics: Euclid, texts, commentators, etc. Paralleling the text is an elaborate critical exposition analyzing each definition, proposition, postulate, etc., and covering textual matters, mathematical analyses, refutations, extensions, etc. Unabridged reproduction of the Cambridge 2nd edition. 3 volumes. Total of 995 figures, 1426pp. 5⅜ x 8.   S88, 89, 90 — 3 vol. set, Paperbound **$6.75**

*THE GEOMETRY OF RENE DESCARTES. The great work which founded analytic geometry. The renowned Smith-Latham translation faced with the original French text containing all of Descartes' own diagrams! Contains: Problems the Construction of Which Requires Only Straight Lines and Circles; On the Nature of Curved Lines; On the Construction of Solid or Supersolid Problems. Notes. Diagrams. 258pp.   S68 Paperbound **$1.60**

*A PHILOSOPHICAL ESSAY ON PROBABILITIES, P. Laplace. Without recourse to any mathematics above grammar school, Laplace develops a philosophically, mathematically and historically classical exposition of the nature of probability: its functions and limitations, operations in practical affairs, calculations in games of chance, insurance, government, astronomy, and countless other fields. New introduction by E. T. Bell. viii + 196pp. S166 Paperbound **$1.35**

DE RE METALLICA, Georgius Agricola. Written over 400 years ago, for 200 years the most authoritative first-hand account of the production of metals, translated in 1912 by former President Herbert Hoover and his wife, and today still one of the most beautiful and fascinating volumes ever produced in the history of science! 12 books, exhaustively annotated, give a wonderfully lucid and vivid picture of the history of mining, selection of sites, types of deposits, excavating pits, sinking shafts, ventilating, pumps, crushing machinery, assaying, smelting, refining metals, making salt, alum, nitre, glass, and many other topics. This definitive edition contains all 289 of the 16th century woodcuts which made the original an artistic masterpiece. It makes a superb gift for geologists, engineers, libraries, artists, historians, and everyone interested in science and early illustrative art. Biographical, historical introductions. Bibliography, survey of ancient authors. Indices. 289 illustrations. 672pp. 6¾ x 10¾. Deluxe library edition.   S6 Clothbound **$10.00**

GEOGRAPHICAL ESSAYS, W. M. Davis. Modern geography and geomorphology rest on the fundamental work of this scientist. His new concepts of earth-processes revolutionized science and his broad interpretation of the scope of geography created a deeper understanding of the interrelation of the landscape and the forces that mold it. This first inexpensive unabridged edition covers theory of geography, methods of advanced geographic teaching, descriptions of geographic areas, analyses of land-shaping processes, and much besides. Not only a factual and historical classic, it is still widely read for its reflections of modern scientific thought. Introduction. 130 figures. Index. vi + 777pp. 5⅜ x 8.
S383 Paperbound **$3.50**

CHARLES BABBAGE AND HIS CALCULATING ENGINES, edited by P. Morrison and E. Morrison. Friend of Darwin, Humboldt, and Laplace, Babbage was a leading pioneer in large-scale mathematical machines and a prophetic herald of modern operational research—true father of Harvard's relay computer Mark I. His Difference Engine and Analytical Engine were the first successful machines in the field. This volume contains a valuable introduction on his life and work; major excerpts from his fascinating autobiography, revealing his eccentric and unusual personality; and extensive selections from "Babbage's Calculating Engines," a compilation of hard-to-find journal articles, both by Babbage and by such eminent contributors as the Countess of Lovelace, L. F. Menabrea, and Dionysius Lardner. 11 illustrations. Appendix of miscellaneous papers. Index. Bibliography. xxxviii + 400pp. 5⅜ x 8.   T12 Paperbound **$2.00**

*THE WORKS OF ARCHIMEDES WITH THE METHOD OF ARCHIMEDES, edited by T. L. Heath. All the known works of the greatest mathematician of antiquity including the recently discovered METHOD OF ARCHIMEDES. This last is the only work we have which shows exactly how early mathematicians discovered their proofs before setting them down in their final perfection. A 186 page study by the eminent scholar Heath discusses Archimedes and the history of Greek mathematics. Bibliography. 563pp. 5⅜ x 8.   S9 Paperbound **$2.45**

Dover Classical Records

Now available directly to the public exclusively from Dover: top-quality recordings of fine classical music for only $2 per record! Originally released by a major company (except for the previously unreleased Gimpel recording of Bach) to sell for $5 and $6, these records were issued under our imprint only after they had passed a severe critical test. We insisted upon:

First-rate music that is enjoyable, musically important and culturally significant.

First-rate performances, where the artists have carried out the composer's intentions, in which the music is alive, vigorous, played with understanding and sympathy.

First-rate sound—clear, sonorous, fully balanced, crackle-free, whir-free.

Have in your home music by major composers, performed by such gifted musicians as Elsner, Gitlis, Wührer, the Barchet Quartet, Gimpel. Enthusiastically received when first released, many of these performances are definitive. The records are not seconds or remainders, but brand new pressings made on pure vinyl from carefully chosen master tapes. "All purpose" 12" monaural 33⅓ rpm records, they play equally well on hi-fi and stereo equipment. Fine music for discriminating music lovers, superlatively played, flawlessly recorded: there is no better way to build your library of recorded classical music at remarkable savings. There are no strings; this is not a come-on, not a club, forcing you to buy records you may not want in order to get a few at a lower price. Buy whatever records you want in any quantity, and never pay more than $2 each. Your obligation ends with your first purchase. And that's when ours begins. Dover's money-back guarantee allows you to return any record for any reason, even if you don't like the music, for a full, immediate refund, no questions asked.

MOZART: STRING QUARTET IN A MAJOR (K.464); STRING QUARTET IN C MAJOR ("DISSONANT", K.465), Barchet Quartet. The final two of the famed Haydn Quartets, high-points in the history of music. The A Major was accepted with delight by Mozart's contemporaries, but the C Major, with its dissonant opening, aroused strong protest. Today, of course, the remarkable resolutions of the dissonances are recognized as major musical achievements. "Beautiful warm playing," MUSICAL AMERICA. "Two of Mozart's loveliest quartets in a distinguished performance," REV. OF RECORDED MUSIC. (Playing time 58 mins.) HCR 5200 **$2.00**

MOZART: QUARTETS IN G MAJOR (K.80); D MAJOR (K.155); G MAJOR (K.156); C MAJOR (K157), Barchet Quartet. The early chamber music of Mozart receives unfortunately little attention. First-rate music of the Italian school, it contains all the lightness and charm that belongs only to the youthful Mozart. This is currently the only separate source for the composer's work of this time period. "Excellent," HIGH FIDELITY. "Filled with sunshine and youthful joy; played with verve, recorded sound live and brilliant," CHRISTIAN SCI. MONITOR. (Playing time 51 mins.) HCR 5201 **$2.00**

MOZART: SERENADE #9 IN D MAJOR ("POSTHORN", K.320); SERENADE #6 IN D MAJOR ("SERENATA NOTTURNA", K.239), Pro Musica Orch. of Stuttgart, under Edouard van Remoortel. For Mozart, the serenade was a highly effective form, since he could bring to it the immediacy and intimacy of chamber music as well as the free fantasy of larger group music. Both these serenades are distinguished by a playful, mischievous quality, a spirit perfectly captured in this fine performance. "A triumph, polished playing from the orchestra," HI FI MUSIC AT HOME. "Sound is rich and resonant, fidelity is wonderful," REV. OF RECORDED MUSIC. (Playing time 51 mins.) HCR 5202 **$2.00**

MOZART: DIVERTIMENTO IN E FLAT MAJOR FOR STRING TRIO (K.563); ADAGIO AND FUGUE IN F MINOR FOR STRING TRIO (K.404a), Kehr Trio. The Divertimento is one of Mozart's most beloved pieces, called by Einstein "the finest, most perfect trio ever heard." It is difficult to imagine a music lover who will not be delighted by it. This is the only recording of the lesser known Adagio and Fugue, written in 1782 and influenced by Bach's Well-Tempered Clavichord. "Extremely beautiful recording, strongly recommended," THE OBSERVER. "Superior to rival editions," HIGH FIDELITY. (Playing time 51 mins.) HCR 5203 **$2.00**

SCHUMANN: KREISLERIANA (OP.16); FANTASY IN C MAJOR ("FANTASIE," OP.17), Vlado Perlemuter, Piano. The vigorous Romantic imagination and the remarkable emotional qualities of Schumann's piano music raise it to special eminence in 19th century creativity. Both these pieces are rooted to the composer's tortuous romance with his future wife, Clara, and both receive brilliant treatment at the hands of Vlado Perlemuter, Paris Conservatory, proclaimed by Alfred Cortot "not only a great virtuoso but also a great musician." "The best Kreisleriana to date," BILLBOARD. (Playing time 55 mins.) HCR 5204 **$2.00**

SCHUMANN: TRIO #1, D MINOR; TRIO #3, G MINOR, Trio di Bolzano. The fiery, romantic, melodic Trio #1, and the dramatic, seldom heard Trio #3 are both movingly played by a fine chamber ensemble. No one personified Romanticism to the general public of the 1840's more than did Robert Schumann, and among his most romantic works are these trios for cello, violin and piano. "Ensemble and overall interpretation leave little to be desired," HIGH FIDELITY. "An especially understanding performance," REV. OF RECORDED MUSIC. (Playing time 54 mins.) HCR 5205 **$2.00**

SCHUMANN: TRIOS #1 IN D MINOR (OPUS 63) AND #3 IN G MINOR (OPUS 110), Trio di Bolzano. The fiery, romantic, melodic Trio #1 and the dramatic, seldom heard Trio #3 are both movingly played by a fine chamber ensemble. No one personified Romanticism to the general public of the 1840's more than did Robert Schumann, and among his most romantic works are these trios for cello, violin and piano. "Ensemble and overall interpretation leave little to be desired," HIGH FIDELITY. "An especially understanding performance," REV. OF RECORDED MUSIC. (Playing time 54 mins.) **HCR 5205 $2.00**

SCHUBERT: QUINTET IN A ("TROUT") (OPUS 114), AND NOCTURNE IN E FLAT (OPUS 148), Friedrich Wührer, Piano and Barchet Quartet. If there is a single piece of chamber music that is a universal favorite, it is probably Schubert's "Trout" Quintet. Delightful melody, harmonic resources, musical exuberance are its characteristics. The Nocturne (played by Wührer, Barchet, and Reimann) is an exquisite piece with a deceptively simple theme and harmony. "The best Trout on the market—Wührer is a fine Viennese-style Schubertian, and his spirit infects the Barchets," ATLANTIC MONTHLY. "Exquisitely recorded," ETUDE. (Playing time 44 mins.) **HCR 5206 $2.00**

SCHUBERT: PIANO SONATAS IN C MINOR AND B (OPUS 147), Friedrich Wührer. Schubert's sonatas retain the structure of the classical form, but delight listeners with romantic freedom and a special melodic richness. The C Minor, one of the Three Grand Sonatas, is a product of the composer's maturity. The B Major was not published until 15 years after his death. "Remarkable interpretation, reproduction of the first rank," DISQUES. "A superb pianist for music like this, musicianship, sweep, power, and an ability to integrate Schubert's measures such as few pianists have had since Schnabel," Harold Schonberg. (Playing time 49 mins.) **HCR 5207 $2.00**

STRAVINSKY: VIOLIN CONCERTO IN D, Ivry Gitlis, Cologne Orchestra; DUO CONCERTANTE, Ivry Gitlis, Violin, Charlotte Zelka, Piano, Cologne Orchestra; JEU DE CARTES, Bamberg Symphony, under Hollreiser. Igor Stravinsky is probably the most important composer of this century, and these three works are among the most significant of his neoclassical period of the 30's. The Violin Concerto is one of the few modern classics. Jeu de Cartes, a ballet score, bubbles with gaiety, color and melodiousness. "Imaginatively played and beautifully recorded," E. T. Canby, HARPERS MAGAZINE. "Gitlis is excellent, Hollreiser beautifully worked out," HIGH FIDELITY. (Playing time 55 mins.) **HCR 5208 $2.00**

GEMINIANI: SIX CONCERTI GROSSI, OPUS 3, Helma Elsner, Harpsichord, Barchet Quartet, Pro Musica Orch. of Stuttgart, under Reinhardt. Francesco Geminiani (1687-1762) has been rediscovered in the same musical exploration that revealed Scarlatti, Vivaldi, and Corelli. In form he is more sophisticated than the earlier Italians, but his music delights modern listeners with its combination of contrapuntal techniques and the full harmonies and rich melodies charcteristic of Italian music. This is the only recording of the six 1733 concerti: D Major, B Flat Minor, E Minor, G Minor, E Minor (bis), and D Minor. "I warmly recommend it, spacious, magnificent, I enjoyed every bar," C. Cudworth, RECORD NEWS. "Works of real charm, recorded with understanding and style," ETUDE. (Playing time 52 mins.) **HCR 5209 $2.00**

MODERN PIANO SONATAS: BARTOK: SONATA FOR PIANO; BLOCH: SONATA FOR PIANO (1935); PROKOFIEV, PIANO SONATA #7 IN B FLAT ("STALINGRAD"); STRAVINSKY: PIANO SONATA (1924), István Nádas, Piano. Shows some of the major forces and directions in modern piano music: Stravinsky's crisp austerity; Bartok's fusion of Hungarian folk motives; incisive diverse rhythms, and driving power; Bloch's distinctive emotional vigor; Prokofiev's brilliance and melodic beauty couched in pre-Romantic forms. "A most interesting documentation of the contemporary piano sonata. Nadas is a very good pianist." HIGH FIDELITY. (Playing time 59 mins.) **HCR 5215 $2.00**

VIVALDI: CONCERTI FOR FLUTE, VIOLIN, BASSOON, AND HARPSICHORD: #8 IN G MINOR, #21 IN F, #27 IN D, #7 IN D; SONATA #1 IN A MINOR, Gastone Tassinari, Renato Giangrandi, Giorgio Semprini, Arlette Eggmann. More than any other Baroque composer, Vivaldi moved the concerto grosso closer to the solo concert we deem standard today. In these concerti he wrote virtuosi music for the solo instruments, allowing each to introduce new material or expand on musical ideas, creating tone colors unusual even for Vivaldi. As a result, this record displays a new area of his genius, offering some of his most brilliant music. Performed by a top-rank European group. (Playing time 45 mins.) **HCR 5216 $2.00**

LÜBECK: CANTATAS: HILF DEINEM VOLK; GOTT, WIE DEIN NAME, Stuttgart Choral Society, Swabian Symphony Orch.; PRELUDES AND FUGUES IN C MINOR AND IN E, Eva Hölderlin, Organ. Vincent Lübeck (1654-1740), contemporary of Bach and Buxtehude, was one of the great figures of the 18th-century North German school. These examples of Lübeck's few surviving works indicate his power and brilliance. Voice and instrument lines in the cantatas are strongly reminiscent of the organ: the preludes and fugues show the influence of Bach and Buxtehude. This is the only recording of the superb cantatas. Text and translation included. "Outstanding record," E. T. Canby, SAT. REVIEW. "Hölderlin's playing is exceptional," AM. RECORD REVIEW. "Will make [Lübeck] many new friends," Philip Miller. (Playing time 37 mins.) **HCR 5217 $2.00**

CATALOGUE OF DOVER BOOKS

DONIZETTI: BETLY (LA CAPANNA SVIZZERA), Soloists of Compagnia del Teatro dell'Opera Comica di Roma, Societa del Quartetto, Rome, Chorus and Orch. Betly, a delightful one-act opera written in 1836, is similar in style and story to one of Donizetti's better-known operas, L'Elisir. Betly is lighthearted and farcical, with bright melodies and a freshness characteristic of the best of Donizetti. Libretto (English and Italian) included. "The chief honors go to Angela Tuccari who sings the title role, and the record is worth having for her alone," M. Rayment, GRAMOPHONE REC. REVIEW. "The interpretation . . . is excellent . . . This is a charming record which we recommend to lovers of little-known works," DISQUES.
HCR 5218 **$2.00**

ROSSINI: L'OCCASIONE FA IL LADRO (IL CAMBIO DELLA VALIGIA), Soloists of Compagnia del Teatro dell'Opera Comica di Roma, Societa del Quartetto, Rome, Chorus and Orch. A charming one-act opera buffa, this is one of the first works of Rossini's maturity, and it is filled with the wit, gaiety and sparkle that make his comic operas second only to Mozart's. Like other Rossini works, L'Occasione makes use of the theme of impersonation and attendant amusing confusions. This is the only recording of this important buffa. Full libretto (English and Italian) included. "A major rebirth, a stylish performance . . . the Roman recording engineers have outdone themselves," H. Weinstock, SAT. REVIEW. (Playing time 53 mins.)
HCR 5219 **$2.00**

DOWLAND: "FIRST BOOKE OF AYRES," Pro Musica Antiqua of Brussels, Safford Cape, Director. This is the first recording to include all 22 of the songs of this great collection, written by John Dowland, one of the most important writers of songs of 16th and 17th century England. The participation of the Brussels Pro Musica under Safford Cape insures scholarly accuracy and musical artistry. "Powerfully expressive and very beautiful," B. Haggin. "The musicianly singers . . . never fall below an impressive standard," Philip Miller. Text included. (Playing time 51 mins.)
HCR 5220 **$2.00**

FRENCH CHANSONS AND DANCES OF THE 16TH CENTURY, Pro Musica Antiqua of Brussels, Safford Cape, Director. A remarkable selection of 26 three- or four-part chansons and delightful dances from the French Golden Age—by such composers as Orlando Lasso, Crecquillon, Claude Gervaise, etc. Text and translation included. "Delightful, well-varied with respect to mood and to vocal and instrumental color," HIGH FIDELITY. "Performed with . . . discrimination and musical taste, full of melodic distinction and harmonic resource," Irving Kolodin. (Playing time 39 mins.)
HCR 5221 **$2.00**

GALUPPI: CONCERTI A QUATRO: #1 IN G MINOR, #2 IN G, #3 IN D, #4 IN C MINOR, #5 IN E FLAT, AND #6 IN B FLAT, Biffoli Quartet. During Baldassare Galuppi's lifetime, his instrumental music was widely renowned, and his contemporaries Mozart and Haydn thought highly of his work. These 6 concerti reflect his great ability; and they are among the most interesting compositions of the period. They are remarkable for their unusual combinations of timbres and for emotional elements that were only then beginning to be introduced into music. Performed by the well-known Biffoli Quartet, this is the only record devoted exclusively to Galuppi. (Playing time 47 mins.)
HCR 5222 **$2.00**

HAYDN: DIVERTIMENTI FOR WIND BAND, IN C; IN F; DIVERTIMENTO A NOVE STROMENTI IN C FOR STRINGS AND WIND INSTRUMENTS, reconstructed by H. C. Robbins Landon, performed by members of Vienna State Opera Orch.; MOZART DIVERTIMENTI IN C, III (K. 187) AND IV (K. 188), Salzburg Wind Ensemble. Robbins Landon discovered Haydn manuscripts in a Benedictine monastery in Lower Austria, edited them and restored their original instrumentation The result is this magnificent record. Two little-known divertimenti by Mozart—of great charm and appeal—are also included. None of this music is available elsewhere (Playing time 58 mins.)
HCR 5223 **$2.00**

PURCELL: TRIO SONATAS FROM "SONATAS OF FOUR PARTS" (1697): #9 IN F ("GOLDEN"), #7 IN C, #1 IN B MINOR, #10 IN D, #4 IN D MINOR, #2 IN E FLAT, AND #8 IN G MINOR, Giorgio Ciompi, and Werner Torkanowsky, Violins, Geo. Koutzen, Cello, and Herman Chessid, Harpsichord. These posthumously-published sonatas show Purcell at his most advanced and mature. They are certainly among the finest musical examples of pre-modern chamber music. Those not familiar with his instrumental music are well-advised to hear these outstanding pieces. "Performance sounds excellent," Harold Schonberg. "Some of the most noble and touching music known to anyone," AMERICAN RECORD GUIDE. (Playing time 58 mins.)
HCR 5224 **$2.00**

BARTOK: VIOLIN CONCERTO; SONATA FOR UNACCOMPANIED VIOLIN, Ivry Gitlis, Pro Musica of Vienna, under Hornstein. Both these works are outstanding examples of Bartok's final period, and they show his powers at their fullest. The Violin Concerto is, in the opinion of many authorities, Bartok's finest work, and the Sonata, his last work, is "a masterpiece" (F. Sackville West). "Wonderful, finest performance of both Bartok works I have ever heard," GRAMOPHONE. "Gitlis makes such potent and musical sense out of these works that I suspect many general music lovers (not otherwise in sympathy with modern music) will discover to their amazement that they like it. Exceptionally good sound," AUDITOR. (Playing time 54 mins.)
HCR 5211 **$2.00**

J. S. BACH: PARTITAS FOR UNACCOMPANIED VIOLIN: #2 in D Minor and #3 in E, Bronislav Gimpel. Bach's works for unaccompanied violin fall within the same area that produced the Brandenburg Concerti, the Orchestral Suites, and the first part of the Well-Tempered Clavichord. The D Minor is considered one of Bach's masterpieces; the E Major is a buoyant work with exceptionally interesting bariolage effects. This is the first release of a truly memorable recording by Bronislav Gimpel, "as a violinist, the equal of the greatest" (P. Leron, in OPERA, Paris). (Playing time 53 mins.) HCR 5212 **$2.00**

ROSSINI: QUARTETS FOR WOODWINDS: #1 IN F, #4 IN B FLAT, #5 IN D, AND #6 IN F, N. Y. Woodwind Quartet Members: S. Baron, Flute, J. Barrows, French Horn; B. Garfield, Bassoon; D. Glazer, Clarinet. Rossini's great genius was centered in the opera, but he also wrote a small amount of first-rate non-vocal music. Among these instrumental works, first place is usually given to the very interesting quartets. Of the three different surviving arrangements, this wind group version is the original, and this is the first recording of these works. "Each member of the group displays wonderful virtuosity when the music calls for it, at other times blending sensitively into the ensemble," HIGH FIDELITY. "Sheer delight," Philip Miller. (Playing time 45 mins.) HCR 5214 **$2.00**

TELEMANN: THE GERMAN FANTASIAS FOR HARPSICHORD (#1-12), Helma Elsner. Until recently, Georg Philip Telemann (1681-1767) was one of the mysteriously neglected great men of music. Recently he has received the attention he deserved. He created music that delights modern listeners with its freshness and originality. These fantasias are free in form and reveal the intricacy of thorough bass music, the harmonic wealth of the "new music," and a distinctive melodic beauty. "This is another blessing of the contemporary LP output. Miss Elsner plays with considerable sensitivity and a great deal of understanding," REV. OF RECORDED MUSIC. "Fine recorded sound," Harold Schonberg. "Recommended warmly, very high quality," DISQUES. (Playing time 50 mins.) HCR 5210 **$2.00**

Nova Recordings

In addition to our reprints of outstanding out-of-print records and American releases of first-rate foreign recordings, we have established our own new records. In order to keep every phase of their production under our own control, we have engaged musicians of world renown to play important music (for the most part unavailable elsewhere), have made use of the finest recording studios in New York, and have produced tapes equal to anything on the market, we believe. The first of these entirely new records are now available.

RAVEL: GASPARD DE LA NUIT, LE TOMBEAU DE COUPERIN, JEUX D'EAU, Beveridge Webster, Piano. Webster studied under Ravel and played his works in European recitals, often with Ravel's personal participation in the program. This record offers examples of the three major periods of Ravel's pianistic work, and is a must for any serious collector or music lover. (Playing time about 50 minutes). Monaural HCR 5213 **$2.00** / Stereo HCR ST 7000 **$2.00**

EIGHTEENTH CENTURY FRENCH FLUTE MUSIC, Jean-Pierre Rampal, Flute, and Robert Veyron-Lacroix, Harpsichord. Contains Concerts Royaux #7 for Flute and Harpsichord in G Minor, Francois Couperin; Sonata dite l'Inconnue in G for Flute and Harpsichord, Michel de la Barre; Sonata #6 in A Minor, Michel Blavet; and Sonata in D Minor, Anne Danican-Philidor. In the opinion of many Rampal is the world's premier flutist. (Playing time about 45 minutes) Monaural HCR 5238 **$2.00** / Stereo HCR ST 7001 **$2.00**

SCHUMANN: NOVELLETTEN (Opus 21), Beveridge Webster, Piano. Brilliantly played in this original recording by one of America's foremost keyboard performers. Connected Romantic pieces. Long a piano favorite. (Playing time about 45 minutes) Monaural HCR 5239 **$2.00** / Stereo HCR ST 7002 **$2.00**

Trubner Colloquial Manuals

These unusual books are members of the famous Trubner series of colloquial manuals. They have been written to provide adults with a sound colloquial knowledge of a foreign language, and are suited for either class use or self-study. Each book is a complete course in itself, with progressive, easy to follow lessons. Phonetics, grammar, and syntax are covered, while hundreds of phrases and idioms, reading texts, exercises, and vocabulary are included. These books are unusual in being neither skimpy nor overdetailed in grammatical matters, and in presenting up-to-date, colloquial, and practical phrase material. Bilingual presentation is stressed, to make thorough self-study easier for the reader.

COLLOQUIAL HINDUSTANI, A. H. Harley, formerly Nizam's Reader in Urdu, U. of London. 30 pages on phonetics and scripts (devanagari & Arabic-Persian) are followed by 29 lessons, including material on English and Arabic-Persian influences. Key to all exercises. Vocabulary. 5 x 7½. 147pp. Clothbound **$1.75**

COLLOQUIAL PERSIAN, L. P. Elwell-Sutton. Best introduction to modern Persian, with 90 page grammatical section followed by conversations, 35-page vocabulary. 139pp.
Clothbound **$2.25**

COLLOQUIAL ARABIC, DeLacy O'Leary. Foremost Islamic scholar covers language of Egypt, Syria, Palestine, & Northern Arabia. Extremely clear coverage of complex Arabic verbs & noun plurals; also cultural aspects of language. Vocabulary. xviii + 192pp. 5 x 7½.
Clothbound **$2.50**

COLLOQUIAL GERMAN, P. F. Doring. Intensive thorough coverage of grammar in easily-followed form. Excellent for brush-up, with hundreds of colloquial phrases. 34 pages of bilingual texts. 224pp. 5 x 7½. Clothbound **$2.00**

COLLOQUIAL SPANISH, W. R. Patterson. Castilian grammar and colloquial language, loaded with bilingual phrases and colloquialisms. Excellent for review or self-study. 164pp. 5 x 7½.
Clothbound **$2.00**

COLLOQUIAL FRENCH, W. R. Patterson. 16th revision of this extremely popular manual. Grammar explained with model clarity, and hundreds of useful expressions and phrases; exercises, reading texts, etc. Appendixes of new and useful words and phrases. 223pp. 5 x 7½.
Clothbound **$2.00**

COLLOQUIAL CZECH, J. Schwarz, former headmaster of Lingua Institute, Prague. Full easily followed coverage of grammar, hundreds of immediately useable phrases, texts. Perhaps the best Czech grammar in print. "An absolutely successful textbook," JOURNAL OF CZECHO-SLOVAK FORCES IN GREAT BRITAIN. 252pp. 5 x 7½. Clothbound **$3.00**

COLLOQUIAL RUMANIAN, G. Nandris, Professor of University of London. Extremely thorough coverage of phonetics, grammar, syntax; also included 70-page reader, and 70-page vocabulary. Probably the best grammar for this increasingly important language. 340pp. 5 x 7½.
Clothbound **$2.75**

COLLOQUIAL ITALIAN, A. L. Hayward. Excellent self-study course in grammar, vocabulary, idioms, and reading. Easy progressive lessons will give a good working knowledge of Italian in the shortest possible time. 5 x 7½. Clothbound **$1.75**

COLLOQUIAL TURKISH, Yusuf Mardin. Very clear, thorough introduction to leading cultural and economic language of Near East. Begins with pronunciation and statement of vowel harmony, then 36 lessons present grammar, graded vocabulary, useful phrases, dialogues, reading, exercises. Key to exercises at rear. Turkish-English vocabulary. All in Roman alphabet. x + 288pp. 4¾ x 7¼. Clothbound **$4.00**

DUTCH-ENGLISH AND ENGLISH-DUTCH DICTIONARY, F. G. Renier. For travel, literary, scientific or business Dutch, you will find this the most convenient, practical and comprehensive dictionary on the market. More than 60,000 entries, shades of meaning, colloquialisms, idioms, compounds and technical terms. Dutch and English strong and irregular verbs. This is the only dictionary in its size and price range that indicates the gender of nouns. New orthography. xvii + 571pp. 5½ x 6¼. T224 Clothbound **$2.75**

LEARN DUTCH, F. G. Renier. This book is the most satisfactory and most easily used grammar of modern Dutch. The student is gradually led from simple lessons in pronunciation, through translation from and into Dutch, and finally to a mastery of spoken and written Dutch. Grammatical principles are clearly explained while a useful, practical vocabulary is introduced in easy exercises and readings. It is used and recommended by the Fulbright Committee in the Netherlands. Phonetic appendices. Over 1200 exercises; Dutch-English, English-Dutch vocabularies. 181pp. 4¼ x 7¼. T441 Clothbound **$2.25**

CATALOGUE OF DOVER BOOKS

INVITATION TO GERMAN POETRY record. Spoken by Lotte Lenya. Edited by Gustave Mathieu, Guy Stern. 42 poems of Walther von der Vogelweide, Goethe, Hölderlin, Heine, Hofmannsthal, George, Werfel, Brecht, other great poets from 13th to middle of 20th century, spoken with superb artistry. Use this set to improve your diction, build vocabulary, improve aural comprehension, learn German literary history, as well as for sheer delight in listening. 165-page book contains full German text of each poem; English translations; biographical, critical information on each poet; textual information; portraits of each poet, many never before available in this country. 1 12″ 33⅓ record; 165-page book; album.   The set **$4.95**

ESSENTIALS OF RUSSIAN record, A von Gronicka, H. Bates-Yakobson. 50 minutes of spoken Russian based on leading grammar will improve comprehension, pronunciation, increase vocabulary painlessly. Complete aural review of phonetics, phonemics—words contrasted to highlight sound differences. Wide range of material: talk between family members, friends; sightseeing; adaptation of Tolstoy's "The Shark;" history of Academy of Sciences; proverbs, epigrams; Pushkin, Lermontov, Fet, Blok, Maikov poems. Conversation passages spoken twice, fast and slow, let you anticipate answers, hear all sounds but understand normal speed. 12″ 33⅓ record, album sleeve. 44-page manual with entire record text. Translation on facing pages, phonetic instructions.   The set **$4.95**

Note: For students wishing to use a grammar as well, set is available with grammar-text on which record is based, Gronicka and Bates-Yakobson's "Essentials of Russian" (400pp., 6 x 9, clothbound; Prentice Hall), an excellent, standard text used in scores of colleges, institutions.   Augmented set: book, record, manual, sleeve **$10.70**

DICTIONARY OF SPOKEN RUSSIAN, English-Russian, Russian-English. Based on phrases and complete sentences, rather than isolated words; recognized as one of the best methods of learning the idiomatic speech of a country. Over 11,500 entries, indexed by single words, with more than 32,000 English and Russian sentences and phrases, in immediately useable form. Probably the largest list ever published. Shows accent changes in conjugation and declension; irregular forms listed in both alphabetical place and under main form of word. 15,000 word introduction covering Russian sounds, writing, grammar, syntax. 15-page appendix of geographical names, money, important signs, given names, foods, special Soviet terms, etc. Travellers, businessmen, students, government employees have found this their best source for Russian expressions. Originally published as War Department Technical Manual TM 30-944. iv + 573pp. 5⅝ x 8⅜.   T496 Paperbound **$3.00**

THE GIFT OF LANGUAGE, M. Schlauch. Formerly titled THE GIFT OF TONGUES, this is a middle-level survey that avoids both superficiality and pedantry. It covers such topics as linguistic families, word histories, grammatical processes in such foreign languages as Aztec, Ewe, and Bantu, semantics, language taboos, and dozens of other fascinating and important topics. Especially interesting is an analysis of the word-coinings of Joyce, Cummings, Stein and others in terms of linguistics. 232 bibliographic notes. Index. viii + 342pp. 5⅜ x 8.   T243 Paperbound **$1.95**

Prices subject to change without notice.

Dover publishes books on art, music, philosophy, literature, languages, history, social sciences, psychology, handcrafts, orientalia, puzzles and entertainments, chess, pets and gardens, books explaining science, intermediate and higher mathematics, mathematical physics, engineering, biological sciences, earth sciences, classics of science, etc. Write to:

Dept. catrr.
Dover Publications, Inc.
180 Varick Street, N.Y. 14, N.Y.